Steam on the CAMBRIAN

Above:
Great Western 'Manor' No 7802 *Bradley Manor* enters Oswestry on the down main with the 4.10pm Whitchurch to Welshpool train, on 8 August 1962. Beyond this two-coach train can be seen Oswestry's engine shed. The line to Gobowen can be seen to the left of the signalbox, and the locomotive works is on the right. *S. D. Wainwright*

Steam on the CAMBRIAN

Rex Kennedy

LONDON

IAN ALLAN LTD

First published 1990

ISBN 0 7110 1891 X

Published by Ian Allan Ltd, Shepperton, Surrey; and printed by Ian Allan Printing Ltd at their works at Coombelands in Runnymede, England

Dedication
To Gisela, a very special lady, and the 'Fleur de Lys'.

Contents

Preface

The Cambrian Railways system has always had a special fascination for me from my introduction to trainspotting in 1943 until the present day. Having lived most of my life in Worcester, holidays as a child were often spent at places such as Barmouth and Aberystwyth, and in BR days my sorties into Mid-Wales involved watching Standards at places such as Caersws, Machynlleth and Towyn. My most vivid memory takes me back to Great Western days when I had the pleasure of seeing double-headed 'Dukes', *Tre Pol and Pen* and *Mounts Bay*, standing at the head of a passenger train at Barmouth.

To me, steam on the Cambrian was special, and I only wished that I had had the opportunity to photograph it in all its glory like others have done. Unfortunately, I did not become a prolific railway photographer until the 1970s. My interest was not only in railways, but in buses too, so please excuse the odd glimpse of a Crosville bus within this album.

Over the years, I have collected many fine photographs and a great deal of detailed information concerning the Cambrian, and at one time I was the proud possessor of the headboard carried by the last 'Cambrian Coast Express' on 4 March 1967.

This book has been compiled in such a way that each section of line relates to the original railway companies that eventually formed the Cambrian Railways Co from 1865. The photographs follow a sequence that enables the reader to follow the route along the line as if travelling by rail over the Cambrian Railways system. The one main omission is the Van Railway, where photographs are few and far between. A reliable source of information once told me that part of the line passed through private land, patrolled by gamekeepers, who did not take kindly to photographers.

My reason for using the English spelling for places such as Towyn and Aberdovey, within this album, is based on the fact that during the greater part of the period covered therein, stations carried names spelt in this way.

I have tried to portray as large a variety of locomotives and rolling stock as possible, showing many locations with their own special points of interest, and special emphasis has been placed on locomotive workings. Within the *Introduction* is a brief history of the lines as they opened, together with detailed information on locomotives and rolling stock over a period of around 100 years. A great deal of this is covered photographically in this book, and will hopefully provide the reader with Cambrian atmosphere. Details of the construction of the Barmouth bridge over the Mawddach estuary has also been given, and accidents, mishaps and bad weather conditions have also been reported upon.

I sincerely hope that the reader will obtain as much pleasure from reading this book, and looking through the nostalgic pictures, as I have had in writing and compiling it.

My grateful thanks go to all those photographers who had the foresight to capture the many wonderful scenes on film, and for giving me permission to include their fine results in this book.

My special thanks go to Mike Polgaze, who has provided a great deal of interesting and unusual information which he has collected over a period of time, and to 'Helly', a young lady who not only had to decipher my sometimes illegible writing, when typing this manuscript, but had to cope with unusually-spelt place names which were foreign to her.

Rex Kennedy
October 1988

Foreword

The Cambrian has always been a rather special railway which has enjoyed great affection from its passengers, its staff and the railway enthusiast fraternity.

To its passengers, it has been both the lifeline which joined otherwise remote communities to each other and to the outside world, and the means by which countless thousands of holidaymakers reached the string of delightful resorts which surround Cardigan Bay and nestle under the mountains of Snowdonia.

To its staff, it has provided employment in an area where jobs have never been plentiful. It has also provided many leaders for the local communities, one excellent railway tradition which still continues to this day. In order to overcome the challenges set by heavy trains, severe gradients and only modestly-sized locomotives, it demanded enginemanship of a high order, another Cambrian tradition. This was reaffirmed in 1987 when trains exceeding the previously published maximum loads for both Class 2 and 'Manor' class locomotives were worked without loss of time.

The affection for the Cambrian amongst those with an interest in railways is demonstrated by the range and number of publications devoted to the memory of this comparatively modest rural railway. Who could fail to be stirred by the sight and sound of a heavy 'Manor'-hauled train being piloted up Talerddig by a 'Dukedog'?

My involvement in the Cambrian started at a very early age. My schoolteacher mother had been evacuated with her class from Birkenhead to Montgomery. After the war, friendship with the farming family who had provided such a welcome in the dark days of 1939 meant return visits to Montgomery. This was reached from our home on Deeside by means of the ex-GCR line from Shotton (High Level) to Wrexham Exchange — thence via Gobowen, Oswestry and Welshpool — or occasionally via Shrewsbury. Later family holidays saw annual pilgrimages to Borth y Gest reached sometimes via Afon Wen or sometimes via Ruabon and Barmouth. As soon as I was old enough to be allowed out on my own, a Saturday trip from Shotton to Wrexham Central provided the opportunity of a trip to Ellesmere. Later Gobowen and Oswestry were investigated and even on one occasion a trip to Llanfyllin was achieved. By this time I had become a regular footplate volunteer on the Ffestiniog Railway and journeys to Portmadoc tended to predominate during the summer months. Although the route via Ruabon and Barmouth was longer than via the North Wales Coast Line it offered more interest, the magnificent view from Barmouth Bridge, and far greater opportunities for illicit footplate riding!

My first digs in Portmadoc were with the family of a guard based at the Cambrian depot. This led to visits to the shed, introductions to the local footplate crews, and a first hand acquaintance with the 'Dukedogs' which created a respect for their power and steaming abilities which has not been dimmed by more recent familiarity with much larger machines.

Subsequent full-time railway employment at Wrexham often brought me back on to Cambrian metals at Dolgelley and Barmouth and the Ffestiniog involvement became a regular summer vacation job whilst at college in the late 1960s. During this period I was reunited with my old friends from Wrexham Rhosddu, the BR Standard Class 3 2-6-2 tanks, which had moved to Machynlleth in 1960 when Rhosddu closed. I am pleased to see that Rex Kennedy has managed to find some good photographs of these much underrated machines. What a pity that one did not find its way to Barry scrapyard and into preservation.

When in 1986 I became the business manager responsible for the Cambrian it was a welcome reunion with an old and much-loved friend. Having seen the success of the Fort William to Mallaig tourist steam service, I was sure that a similar project on the Cambrian Coast line could be equally successful. I still believe that the view from a train on Barmouth Bridge is without equal in Britain. With the active support of the Wales Tourist Board a detailed scheme was developed and agreed for the 1987 summer season. Despite a summer of very mixed weather when the heavens seemed to open every Sunday without fail, the service loaded reasonably well and a quite respectable profit was declared. In addition, the Sprinters had carried a significant increase in business in which the publicity generated by the steam trains must have been a major factor.

Unfortunately a reorganisation in the management of railways in Wales in 1988 ended my responsibility for the Cambrian lines. Without anyone else ready and willing to take over the not insignificant task of organising and developing the steam service it failed to reappear in 1988. This was a great pity as much had been achieved and the tourist trade had enjoyed a considerable boost. At least we proved it could be done, and the possibility remains that at some future date it could be resurrected. I am sure that the history of Steam on the Cambrian has not yet reached its final chapter and still hope that one day we might again see a 'Dukedog' back home again on the line for which they were designed.

David McIntosh
Provincial Manager East & West Midlands
(Midlands and Mid Wales 1986-88)

Introduction

History

Steam on the Cambrian commenced in 1859 and ceased in 1967, with the exception of the Vale of Rheidol Railway. The bill to construct the Llanidloes & Newtown Railway (L&NR) received the Royal Assent in 1853, and this was the first of the Cambrian lines to open to traffic, in 1859. This was followed in 1861 by the opening of the Oswestry & Newtown Railway (O&NR), together with a branch line to Porthywaen, a line later extended to Blodwell Junction and the Tanat Valley Light Railway (TVLR). In 1863, the Kerry and Llanfyllin branches opened to traffic, and although the Newtown & Machynlleth Railway (N&MR) was opened on 1 May 1862, it was not officially opened to passenger traffic until 3 January 1863.

The next section of the Cambrian Railways system to be completed was the Oswestry, Ellesmere & Whitchurch Railway, on 27 July 1864, followed by the coast route to Aberystwyth on 1 August 1864. The Cambrian Railways Co, therefore, at its inception on 25 July 1864, being formed of all the above-named companies, could boast almost 100 miles of main line from Whitchurch to Aberystwyth. A further 53½ miles was added in October 1867 when the Aberystwyth & Welsh Coast Railway opened to Pwllheli. The coast lines were now complete, and in 1869 the line from Barmouth Junction to the Great Western Railway's Dolgelley station was opened.

The Van Railway, promoted privately by Earl Vane and David Davies, opened to the Van lead mines for freight in 1871, and to passengers in December 1873. This closed in 1893 after a slump in the price of lead.

However, I do have in my possession interesting documentation dated 17 May 1923, relating to trials carried out on the Van branch, involving the use of the Newton station shunting horse, in order to ascertain if it was possible for traffic to be horse-drawn, hoping that the running of a locomotive could be dispensed with.

Two trials were carried out with a gross load of 14tons 11cwt — one full truckload of coal. The Van branch gradient from Garth Road was 1 in 85, and the horse walked between the rails, but it was only able to pull the wagon 120yd up the gradient to a point more than a quarter of a mile from the terminus. However, hauling a gross weight of 7tons 15cwt, the horse could pull the load 20yd further.

Apart from the 1 in 95 incline at Cerist, this gradient from Garth Road was the only real obstacle of this nature on the line. It was decided that it was unreasonable and impracticable to expect a horse to cover the entire 6-mile length of the branch.

On 16 June 1923 another trial run was made from Caersws to Van using two shunting horses. The three river bridges on the line were planked for use by the horses, and the trial was successful as they were able to haul a two-wagon gross load of 28tons 9cwt as far as 150yd short of Cerist station, from where it became necessary to pull one wagon at a time over the 1 in 95 set of curves for a further 200yd. From here the load was again reunited, but after returning to double-heading, it proved too much, so the brake van was detached.

The experiment proved that horse haulage was only possible to Trefeglwys with two horses and two loaded wagons — beyond here two horses could only haul 13 tons. The entire trial run from Caersws to Vanhead with two horses took 3hr 39min over the 6½-mile 'course'. Traffic generally conveyed over the Van branch included coal, flour, general groceries, etc.

The report stated that if the horse-drawn option was taken up, three animals would be required (one for rest and relay purposes) and a horse 'stabling point' would be required at Caersws. The horses used in the trial were Nos 925 and 938, and a great deal more information appears within the documentation — but these are the general facts.

In April 1888, the Cambrian took over workings on the independent Mid-Wales Railway from Llanidloes to Talyllyn Junction, but final amalgamation did not take place until July 1904.

Final additions to the Cambrian system were the Nantmawr branch in February 1881, the Wrexham & Ellesmere Railway (W&ER) in November 1895, the Van Railway in July 1896 (reopened), and the Tanat Valley Light Railway being incorporated into the Cambrian in March 1921, although opening as an independent company in January 1904. The Mawddwy Railway, also originally an independent company, was reopened by the Cambrian in 1910, having first opened in October 1867, and closed soon after.

The two narrow gauge railways, the Vale of Rheidol and the Welshpool & Llanfair, opened to passenger traffic on 22 December 1902, and 4 April 1903 respectively. The Vale of Rheidol was eventually amalgamated into the Cambrian in July 1913. These companies went to form the Cambrian Railways, making a system of approximately 300 miles which was taken over by the GWR in 1922, and was the largest railway company in Wales.

David Davies and Thomas Savin were responsible for the construction of the lines on the Cambrian system and provided locomotives and rolling stock. They also operated the railways, but the partnership was dissolved in October 1860 with David Davies completing the N&MR, and Thomas Savin the O&NR, and taking on the A&WCR. He also constructed the Oswestry, Ellesmere & Whitchurch Railway. Davies and Savin found new partners but, in February 1866, Thomas Savin went bankrupt, and the Cambrian Railways took over control completely. Prior to that, Savin had taken over the L&NR and the O&NR from Davies and the joint L&N and

The Cambrian Railways System in 1922.

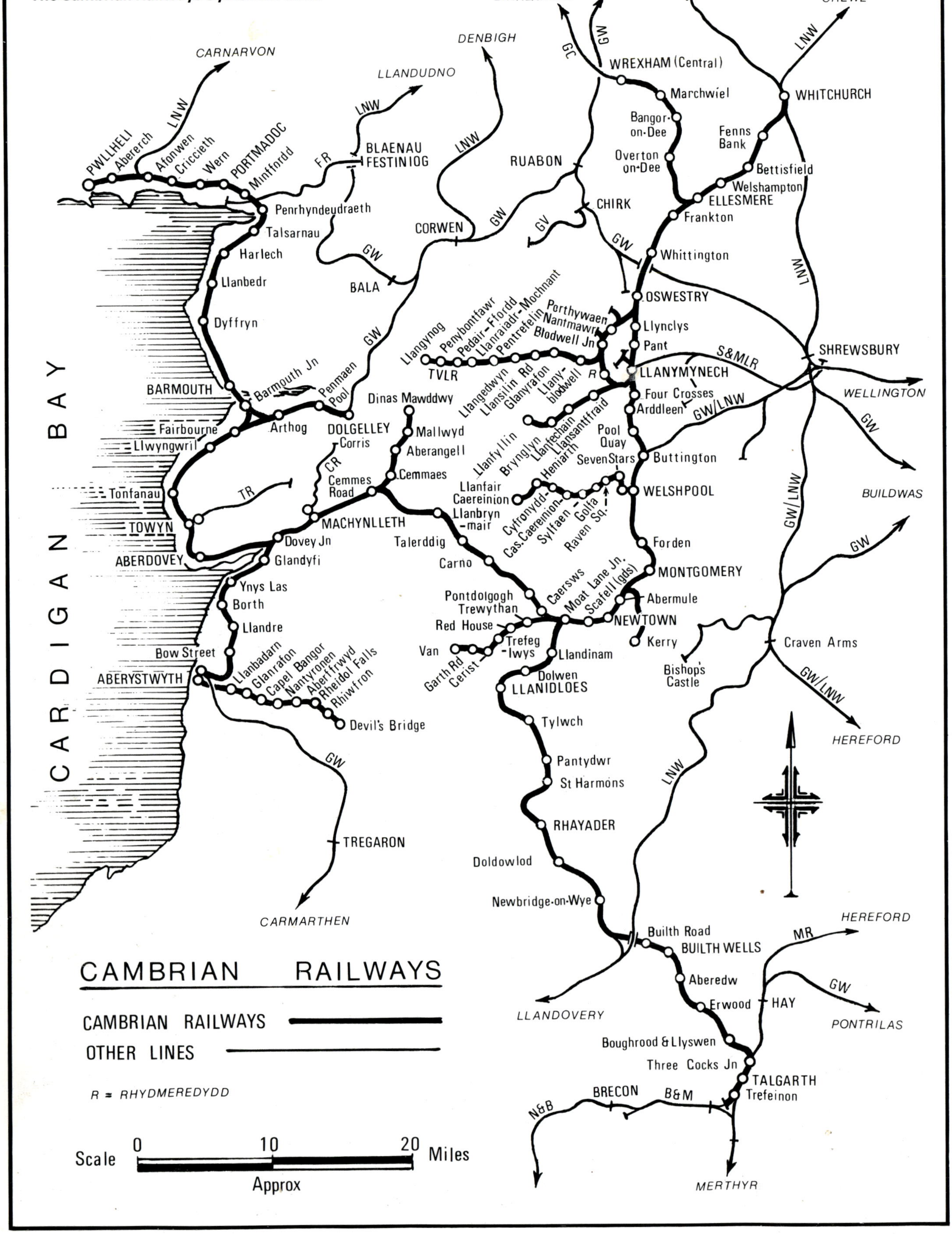

O&N Committee paid Savin two-thirds of the purchase price of all locomotives, the balance being mortgaged.

In 1866, Oswestry Works opened, construction having commenced in 1864. This replaced a small workshop situated at Welshpool.

Locomotives

During the Savin period, Sharp Stewart 0-6-0s with inside cylinders and 4ft 6in wheels were the standard freight locomotives. Savin purchased 16 of these locomotives and all were named – 13 of these became Cambrian property. These were all built between 1861 and 1865, those completed prior to 1864 being fitted with four-wheel tenders, and the four built in 1864/5 (Nos 45, 46, 51 and 52) having six-wheeled tenders. These four locomotives, together with No 40 *Cyfronydd* (with a four-wheeled tender), were withdrawn by the GWR, No 45 *Rhiewport* as late as September 1945. Savin ordered 2-4-0s from Sharp Stewart for passenger work on the lines which he was constructing, and 12 of these eventually became Cambrian locomotives. These were also named and six of these were withdrawn by the GWR – the last, No 55 *Treflach* in November 1922.

These two particular classes have been singled out as they contain locomotives which lasted to the Grouping and passed to the Great Western Railway. A full list of Cambrian engines absorbed into the GWR is given overleaf, showing their Great Western numbers. Many of the final class of 15 0-6-0s built primarily for goods and excursion traffic lasted until the early/mid-1950s. Five were built by Robert Stephenson & Co in 1903, five by Beyer Peacock in 1908, with a further five from this manufacturer in 1918/19. They were the first Cambrian locomotives to be fitted with Belpaire fireboxes.

Left:
Sharp Stewart 0-6-0 No 27 (originally named *Cambria*) never passed to the GWR as it was withdrawn in December 1919. Built for Savin in 1863, and used on the Oswestry & Newtown Railway, it is pictured at Machynlleth in June 1909. These locomotives were built for freight work and had 4ft 6in driving wheels.
H. L. Hopwood

Below:
Another locomotive which was withdrawn prior to the Grouping was 2-4-0 No 28 (originally named *Mazeppa*), built by Sharp Stewart for Savin in 1863, with 5ft 6in driving wheels, and also used on the Oswestry & Newtown Railway. Its life ended on the Cambrian in December 1920 and was one of a class of 12 locomotives built for passenger work. It is seen at Machynlleth in July 1909.
H. L. Hopwood

Cambrian locomotives absorbed into the GWR

Cambrian No	GWR No	Manufacturer	Wheel Arrangement	Date Built	Date Withdrawn
1	1329	Wolverhampton (GWR)	2-4-0	1883	11/1927
2	1129	Beyer Peacock	4-4-0T	1864	9/1922
3	10	Nasmyth Wilson	0-4-4T	1895	10/1932
4	897	Sharp Stewart	0-6-0	1872	7/1922
5	11	Nasmyth Wilson	0-4-4T	1895	5/1922
7	15	Nasmyth Wilson	0-4-4T	1895	10/1922
8	19	Nasmyth Wilson	0-4-4T	1899	10/1932
9	20	Nasmyth Wilson	0-4-4T	1899	6/1928
10	1328	Wolverhampton (GWR)	2-4-0	1883	3/1926
11	1068	Oswestry (Cam Rlys)	4-4-0	1904	1/1924
12	1130	Beyer Peacock	4-4-0T	1864	6/1923
14	898	Sharp Stewart	0-6-0	1875	11/1947
15	844	Beyer Peacock	0-6-0	1918	8/1954
16	1115	Sharp Stewart	4-4-0	1878	1/1925
17	1116	Sharp Stewart	4-4-0	1878	9/1924
19	1082	Oswestry (Cam Rlys)	4-4-0	1901	4/1928
20	1117	Sharp Stewart	4-4-0	1886	5/1922
21	1118	Sharp Stewart	4-4-0	1886	10/1930
23	21	Nasmyth Wilson	0-4-4T	1899	5/1922
24	819	Hunslet	0-6-0T	1903	3/1946
26	820	Chapman & Furneaux	0-6-0T	1898	3/1930 (sold)
29	849	Beyer Peacock	0-6-0	1918	10/1954
30	824	Manning Wardle	0-6-0T	1865	9/1940
31	855	Beyer Peacock	0-6-0	1919	10/1954
32	1085	R. Stephenson	4-4-0	1897	3/1923
33	1131	Beyer Peacock	4-4-0T	1864	7/1922
34	1113	Beyer Peacock	4-4-0	1864	5/1922
35	821	Chapman & Furneaux	0-6-0T	1898	10/1932 (sold)
36	1114	Beyer Peacock	4-4-0	1864	5/1922
37	1132	Beyer Peacock	4-4-0T	1885	1/1923
38	864	Beyer Peacock	0-6-0	1908	11/1952
40	899	Sharp Stewart	0-6-0	1863	7/1922
41	1330	Sharp Stewart	2-4-0	1864	7/1922
42	873	Beyer Peacock	0-6-0	1919	3/1954
43	1331	Sharp Stewart	2-4-0	1864	7/1922
44	1190	Sharp Stewart	2-4-0T	1864	7/1922

Left:
Of the six ex-Metropolitan Railway 4-4-0Ts which were purchased by the Cambrian in 1905, No 34, pictured here at Aberystwyth in June 1909, was one of the two 4-4-0Ts which were converted to tender engines in 1915. The conversion was carried out by Beyer Peacock of Manchester. Driving wheels were 5ft 10in and the Prince of Wales's feathers and 'CAMBRIAN RAILWAYS' on the tank side were added at Oswestry when the Metropolitan condensing gear was removed. Note the raised brass numerals on the coal bunker. Whilst in Metropolitan ownership, this engine carried the name *Daphne*. *H. L. Hopwood*

Right:
No 56, a passenger engine built for the Cambrian by Sharp Stewart in 1865 was, in May 1907, converted from a 2-4-0 tender engine to a 2-4-0T. It is pictured here at Machynlleth in July 1909 in its converted state. When new, it was named *Whittington*. Note the raised brass numerals on the coal bunker. No 56 was allocated the number 1197 by the GWR, but was withdrawn in October 1922. *H. L. Hopwood*

Cambrian No	GWR No	Manufacturer	Wheel Arrangement	Date Built	Date Withdrawn
45	900	Sharp Stewart	0-6-0	1864	9/1945
46	901	Sharp Stewart	0-6-0	1864	7/1922
47	1086	R. Stephenson	4-4-0	1897	5/1923
48	908	Sharp Stewart	0-6-0	1873	12/1938
49	909	Sharp Stewart	0-6-0	1873	7/1922
50	1110	Sharp Stewart	4-4-0	1891	1/1925
51	910	Sharp Stewart	0-6-0	1865	5/1935
52	911	Sharp Stewart	0-6-0	1865	7/1922
53	1332	Sharp Stewart	2-4-0	1865	8/1922
54	874	Beyer Peacock	0-6-0	1919	3/1933
55	1333	Sharp Stewart	2-4-0	1865	11/1922
56	1191	Sharp Stewart	2-4-0T	1865	10/1922
57	1192	Sharp Stewart	2-4-0T	1866	8/1929
58	1196	Sharp Stewart	2-4-0T	1866	4/1948
59	1197	Sharp Stewart	2-4-0T	1866	4/1948
60	1112	Sharp Stewart	4-4-0	1891	6/1928
61	1088	Sharp Stewart	4-4-0	1893	3/1926
62	1090	Sharp Stewart	4-4-0	1893	9/1928
63	1091	Sharp Stewart	4-4-0	1893	10/1930
64	1093	Sharp Stewart	4-4-0	1893	4/1926
65	1096	Sharp Stewart	4-4-0	1893	11/1922
66	1097	Sharp Stewart	4-4-0	1893	11/1927
67	1100	Sharp Stewart	4-4-0	1893	7/1930
68	1101	Sharp Stewart	4-4-0	1893	10/1922
69	1102	Sharp Stewart	4-4-0	1894	12/1925
70	1103	Sharp Stewart	4-4-0	1894	7/1922
71	1104	Sharp Stewart	4-4-0	1894	9/1925
72	1105	Sharp Stewart	4-4-0	1894	7/1922
73	875	Neilson	0-6-0	1894	12/1936
74	876	Neilson	0-6-0	1894	5/1946
75	878	Neilson	0-6-0	1894	10/1938
76	879	Neilson	0-6-0	1894	5/1932
77	880	Neilson	0-6-0	1894	11/1938
78	881	Vulcan Foundry	0-6-0	1895	7/1932
79	882	Vulcan Foundry	0-6-0	1895	1/1935

Cambrian No	GWR No	Manufacturer	Wheel Arrangement	Date Built	Date Withdrawn
80	883	Vulcan Foundry	0-6-0	1895	11/1926
81	1084	Sharp Stewart	4-4-0	1895	5/1922
82	3521	Swindon (GWR)	4-4-0	1887	10/1931
83	1106	Sharp Stewart	4-4-0	1895	4/1931
84	1107	Sharp Stewart	4-4-0	1895	5/1925
85	1108	R. Stephenson	4-4-0	1898	11/1927
86	1109	R. Stephenson	4-4-0	1898	1/1924
87	884	Neilson Reid	0-6-0	1899	8/1947
88	885	Neilson Reid	0-6-0	1899	8/1944
89	887	R. Stephenson	0-6-0	1903	11/1952
90	888	R. Stephenson	0-6-0	1903	5/1922
91	889	R. Stephenson	0-6-0	1903	8/1922
92	891	R. Stephenson	0-6-0	1903	8/1922
93	892	R. Stephenson	0-6-0	1903	4/1953
94	1014	R. Stephenson	4-4-0	1904	4/1928
95	3546	Swindon (GWR)	4-4-0	1888	3/1927
96	1029	R. Stephenson	4-4-0	1904	4/1928
97	1035	R. Stephenson	4-4-0	1904	5/1928
98	1043	R. Stephenson	4-4-0	1904	1/1933
99	893	Beyer Peacock	0-6-0	1908	2/1953
100	894	Beyer Peacock	0-6-0	1908	4/1953
101	895	Beyer Peacock	0-6-0	1908	10/1954
102	896	Beyer Peacock	0-6-0	1908	4/1953
Vale of Rheidol locomotives					
1	1212	Davies & Metcalfe	2-6-2T	1902	12/1932
2	1213	Davies & Metcalfe	2-6-2T	1902	7/1924
3	1198	Bagnall	2-4-0T	1896	7/1923
Welshpool & Llanfair locomotives					
1	822	Beyer Peacock	0-6-0T	1902	8/1961 (sold)
2	823	Beyer Peacock	0-6-0T	1902	7/1962 (sold)

Notes:

GWR locomotives Nos 11/15/21/889/891/897/899/901/909/911/1084/1096/1101/1103/1105/1113/1114/1117/1129/1131/1132/1190/1191/1330/1331/1332/1333 never carried their GWR numbers, and Cambrian engines Nos 82/95 did not either.

Of the six ex-Metropolitan Railway engines (Cambrian numbers 2/12/33/34/36/37), Nos 34 and 36 were converted from 4-4-0Ts to 4-4-0s in 1915/16, and Cambrian engines Nos 44 and 46 were converted from 2-4-0 to 2-4-0T in 1907.

GWR No 1106 was renumbered 1110 in 1926, VoR No 1213 was renumbered 9 in 1949, and Cambrian engines Nos 99-102 ran as Nos 15/31/42/54 until October 1908.

GWR engines Nos 3521 and 3546 were converted from tank locomotives in 1899 and 1900.

In November 1905, the Cambrian Railways purchased six Beyer Peacock 4-4-0Ts from the Metropolitan Railway, which became redundant after electrification of that line. These were found to be unsuitable for branch line work, and when tried on banking duties from Machynlleth to Talerddig, their tank and bunker capacity was found to be insufficient. As no suitable work could be found for them, they were used on shunting duties and on short passenger trips from Oswestry, Machynlleth and Aberystwyth. On purchase from the Metropolitan, at a bargain price of £500 each, their condensing gear was removed at Oswestry and they were painted in Cambrian green livery with the words 'Cambrian Railways' and the Prince of Wales's feathers on the tank sides. The new Cambrian number was in raised brass and placed on the bunker. Nos 34 and 36 (formerly Metropolitan Railway Nos 13 and 15) were converted to tender engines. All were allocated GWR numbers at the Grouping but only No 1130 (ex-Cambrian No 12) ever carried the GWR number. This locomotive was the last of these six locomotives to be withdrawn, in June 1923.

After the Grouping, a great deal of the Cambrian engines were sent to Swindon and never returned, and replacements were required to operate the Cambrian lines. As only the smaller GWR tender engines could be used, this placed many restrictions as to which locomotives could take over. These included the 'Stellas', 'Barnums', 'Dean Goods', 3232 class and eventually, the 'Dukes'.

The most suitable seemed to be the 4-4-0 'Dukes' with Nos 3261, 3269, 3270 and 3271 being the first of this class to be sent to Wales. Eleven 'Dukes' were allocated to Cambrian depots by 1926 and, from 1929-35, up to 20 could be found on the system. The 'Dukes' gradually replaced 'Barnums', 10 of which were sent to the Cambrian for passenger work and double-heading in 1922. These 2-4-0s with their 6ft 2in coupled wheels were allocated to Oswestry and Machynlleth, and the last to be withdrawn was No 3213 in June 1936. The

Left:
From 1878 to 1891, six 4-4-0 passenger locomotives were built by Sharp Stewart, as the earlier 2-4-0s has difficulty in coping with the heavier passenger trains during the summer months. No 60 was built in 1891 and became GWR No 1112. It was withdrawn in June 1928 after being fitted with a replacement boiler in November 1914.
H. L. Hopwood

Left:
No 78 was a Vulcan Foundry-built 0-6-0 of the 73 class of large goods engines, and was constructed in 1895. This locomotive eventually carried the GWR number 881, and was withdrawn in July 1932. Prior to passing to the GWR, No 78 received a reconditioned boiler. A superheated boiler was fitted to No 881 in May 1924 at Swindon. Six others in this class were treated accordingly — one at Swindon and five at Oswestry. No 78 is pictured at Machynlleth in June 1909. Note the oval numberplate. *H. L. Hopwood*

Cambrian also received three members of the 3232 class and seven 'Stella' class locomotives. These operated from Oswestry and Machynlleth, and some from Brecon, but all had disappeared by late 1930.

Swindon-built 4-4-0s, Nos 3521 and 3546 of 1887/88, on the Cambrian since 1921, were joined by six others of the same class for use mainly in the Machynlleth and Aberystwyth area. All were withdrawn by 1931. The original two were purchased to replace Nos 82 and 95, destroyed at Abermule.

The 'Dean Goods' locomotives worked passenger and freight trains from 1922, and these could be seen well into the 1950s.

Other imports included 2-4-0T No 1308 *Lady Margaret* (ex-Liskeard & Looe Railway), which made its home at Oswestry from 1922 to withdrawal in 1948, two 0-6-0ST Pecketts, Nos 679 and 680 (ex-Alexandra Docks Railway), former Whitland & Cardigan 0-6-0ST No 1331, and No 1376, a 0-6-0T from the Bristol & Exeter Railway. These were used for freight duties on the Porthywaen and Tanat Valley lines.

The 'Dukedogs' arrived from 1936 onwards and were numbered 3200-3228 (later 9000-28).

Vale of Rheidol 1ft 11½in gauge engines, built for the opening of the line in 1902, were constructed by Davies & Metcalfe of Romiley. They were 2-6-2 tank engines and numbered 1, *Edward VII* and 2, *Prince of Wales*, changed to 1212 and 1213 respectively by the GWR at the Grouping. The third locomotive acquired for the opening of the line was 2-4-0T No 3 *Rheidol*, built in 1896 by Bagnalls.

The original Cambrian livery was black, with one word, 'Cambrian', in block capitals on the tanksides. No 1212 was sent to Swindon in 1932 and cut-up in March 1935. In 1923 Swindon built two more 2-6-2Ts for the VoR, Nos 7 and 8, *Owain Glyndwr* and *Llywelyn*. They did not receive their names until 1956. No 1213 was 'renewed' at Swindon in 1924 — for accountancy purposes it was a rebuild, but in reality it was totally new. It was numbered 9.

The two Welshpool & Llanfair Light Railway locomotives,

No 1, *The Earl* and No 2, *The Countess* (later 822/823) were not strictly Cambrian engines. They were built by Beyer Peacock in 1902 for the 2ft 6in gauge line which opened in April 1903. Passenger traffic on the Welshpool & Llanfair Railway (W&L) ceased in February 1931, but freight continued to November 1956, when the locomotives were put in store at Oswestry. They were sold to the Welshpool & Llanfair Light Railway Preservation Society in 1961 and 1962.

A New Era

By 1948, only three 4-6-0 'Manors' were allocated to Oswestry (Nos 7807/08/19) and only two to Machynlleth (Nos 7802/03) — the latter including Aberystwyth. A typical 1948 allocation is listed below:

Machynlleth (including Aberystwyth and VoR)

Ex-Corris Railway (not Cambrian stock)	Nos 3/4 (2ft 3in gauge)
Vale of Rheidol Railway, 2-6-2T	Nos 7/8/1213 (1ft 11½in gauge)
Ex-Cambrian locomotives, 0-6-0	Nos 864/892/894
1400 class, 0-4-2T	Nos 1465/74
850 class, 0-6-0PT	No 1965
2021 class, 0-6-0PT	No 2151
2251 class, 0-6-0	Nos 2200/04/19/23/83/98/3200/01/07
'Dean Goods', 0-6-0	Nos 2323/45/56/2464/2572
4500 class, 2-6-2T	Nos 4501/11/13/49/55/60/71/75/5517/24/41/60/70
7400 class, 0-6-0PT	Nos 7406/10
'Manor' class, 4-6-0	Nos 7802/03
'Dukedog', 4-4-0	Nos 9002/04/05/09/12/13/14/17/21/25/27
'Duke', 4-4-0	Nos 9054/72/89/91

Total Allocation: 58 locomotives

(Machynlleth also supplied engines to Penmaenpool and Aberayron.)

Oswestry

Ex-Alexandra Docks & Railway 0-6-0STOC	No 680
Welshpool & Llanfair Light Railway	Nos 822/823 (2ft 6in gauge)
Ex-Cambrian locomotives, 0-6-0	Nos 844/49/55/73/76/84/87/93/95/96/98
Ex-Cambrian locomotives, 2-4-0T	Nos 1196/1197
Ex-Liskeard & Looe Railway, 2-4-0T	No 1308
Ex-Whitland & Cardigan Railway, 0-6-0ST	No 1331
1400 class, 0-4-2T	Nos 1412/17/59/5806
2021 class, 0-6-0PT	Nos 2032/54/68/75
2251 class, 0-6-0	Nos 2201/10/44/55/60/3203/08
'Dean Goods', 0-6-0	Nos 2327/54/82/86/449/52/82/83/516/43/56
7400 class, 0-6-0PT	No 7406
'Manor' class, 4-6-0	Nos 7807/08/19
8100 class, 2-6-2T	No 8103
'Dukedog', 4-4-0	Nos 9000/01/03/16/20/22/26/28
'Duke', 4-4-0	No 9065

Total Allocation: 58 locomotives

Below:
Also sporting an oval numberplate, No 83 was a 61 class passenger locomotive, built in 1895 by Sharp Stewart, and withdrawn from service in 1931. On passing to the GWR it was renumbered 1106, but was again renumbered to 1110 in February 1926. No 83 (together with No 84) was fitted with Davies & Metcalfe exhaust injectors, which became standard equipment on all future Cambrian classes owing to their success. Of the 22 locomotives in this class, 16 were built by Sharp Stewart, four by Robert Stephenson & Co, and two at Oswestry Works. *H. L. Hopwood*

Above:
Large Belpaire class passenger 4-4-0 No 95 was one of the two locomotives (with No 82) involved in the tragic head-on collision near Abermule on 26 January 1921 (*see 'Accidents'*). It was one of five built in 1904 by Robert Stephenson & Co. These locomotives had 6ft driving wheels and, together with their tenders, weighed 77 tons. No 95 is seen at Aberystwyth in June 1909 sporting its oval numberplate, and the circular version of the Cambrian crest is seen on the splasher.
H. L. Hopwood

As can be seen from the 1948 allocations, both Oswestry and Machynlleth had an allocation of 58 locomotives. This compares with the 1938 allocations of Oswestry (69) and Machynlleth (29). However, in 1938, the Oswestry allocation included sub-shed deployment of Llanfyllin (1), Llanidloes (4), Moat Lane (7), Welshpool (1), W&L (2) and Whitchurch (1). The Machynlleth allocation included locomotives sub-shedded at Aberystwyth (21), VoR (3), Aberayron (1), Corris (2), Portmadoc (6) and Pwllheli (3). Also, eight of the Oswestry allocation were at Brecon.

The first 4-6-0 locomotive to work over the Cambrian was No 7808 *Cookham Manor* (now preserved) — in 1943. The 'Manors' were gradually moved from South Devon with the introduction of the diesel-hydraulics there and, between 1947 and 1960, the allocation of 'Manors' on the Cambrian changed from three at Oswestry and two at Machynlleth to eight at Oswestry and six at Machynlleth. In the Machynlleth allocation will be seen a predominance of 4500 class 2-6-2Ts. No 5570 was the first of this class to be shedded there, in 1936, and in 1937 it was joined by No 4549. By 1947 there were 15 of the class allocated to Machynlleth. No 8103 (2-6-2T) was allocated to Oswestry from 1947 to 1957.

During World War 2, the following LMS locomotives were based in Mid-Wales, mainly at Oswestry, on a 'loan' basis. All were 0-6-0s, and the table below details their movements to and from the Cambrian:

No	To stock	Returned to LMS
3038	2/40	11/45
3039	12/39	11/45
3047	10/39	11/45
3085	12/39	11/45
3126	12/39	11/45
3196	12/39	11/45
3536	12/39	8/45
3689	11/39	11/45
3739	2/40	11/45

On 14 March 1944, a 'Special' notice was sent by the GWR's Chief Engineer's Office at Aldermaston to the Divisional Engineer's Office at Oswestry regarding the passage of 'overseas' ambulance trains. This notice included a diagram, together with dimensions of an LNER B12 4-6-0 locomotive, allocated to haul these trains. A list of routes were included in the notice, the Cambrian ones being Whitchurch to Oswestry and Oswestry to Gobowen — to be used by Bristol to Birkenhead, via Severn Tunnel and Shrewsbury, trains. The B12s were permitted to work double-headed with a GWR locomotive over the designated Cambrian routes providing all engines worked chimney-first.

The first 2251 Class 0-6-0s appeared on the Cambrian in 1938 to replace the 'Dean Goods' locomotives, and in 1954 the allocation was Oswestry (5) and Machynlleth (9).

Above:
Between 1903 and 1919, the large Belpaire 15 class goods engines were constructed by Robert Stephenson & Co (five engines) and Beyer Peacock (10 engines). GWR No 873 (ex-Cambrian No 42) was built in 1919 by Beyer Peacock and lasted until March 1954. It is seen at Oswestry shed in this view and, in fact, No 873 recorded the highest Great Western mileage of any Cambrian engine — 798,614. These were the first Cambrian engines to be fitted with Belpaire fireboxes and they were 'yellow' group locomotives. *E. E. Smith*

By 1959, the Standards and ex-LMS Moguls had arrived on the scene, and the 'Dean Goods' and 'Dukedogs' had disappeared. Around 1963 the allocation of Cambrian-based locomotives looked like this:

Oswestry

2251 class, 0-6-0	Nos 2233/3208/09
5700 class, 0-6-0PT	Nos 3600/3770/89/9681
5400 class, 0-6-0PT	No 5422
7400 class, 0-6-0PT	No 7410
4300 class, Mogul, 2-6-0	No 6342
'Manor' class, 4-6-0	Nos 7800/01/07/09/10/19/22
Ivatt Mogul, 2-6-0	Nos 46401/503/04/05/07/08/09/10/12/13/14/15/18/21/22/23/24/25/26/27

Total allocation: 37 locomotives

Machynlleth

2251 class, 0-6-0	Nos 2255/86/94
4500 class, 2-6-2T	Nos 5510/40/53/65
4300 class, Mogul, 2-6-0	Nos 6371/92
7400 class, 0-6-0PT	No 7434
Stanier 3MT, 2-6-2T	Nos 40085/86
Class 4 Standard, 4-6-0	Nos 75020/26
Class 2 Standard, 2-6-0	Nos 78000/02/03/06
Class 3 Standard, 2-6-2T	Nos 82000/20/21/31

Total allocation: 22 locomotives

Aberystwyth

2251 class, 0-6-0	Nos 2217/60/98
7400 class, 0-6-0PT	No 7417
'Manor' class, 4-6-0	Nos 7802/03/14/15/18
Vale of Rheidol, 2-6-2T	Nos 7/8/9

Total allocation: 12 locomotives

It is noticeable that, at this period, Aberystwyth had acquired its own allocation. Also, for the record, at this particular time, locomotives Nos 2204, 7405, 7827, 75005 and 78005, allocated to the Central Wales Division, were receiving works attention at Oswestry, Wolverhampton (Stafford Road) and Caerphilly.

The Cambrian system was taken over by the London Midland Region on 1 January 1964.

The variety of motive power over the Cambrian system has been a great attraction to enthusiasts, coupled with the fine rural scenery which enhances this part of Britain's railway network. Since the war it has seen ex-Cambrian locomotives, 'Dukes' and 'Dukedogs', GWR side tanks, pannier tanks, the infamous 'Manors', LMS Ivatt 2-6-0s, Stanier tanks and the inevitable Standards.

Constant reference has been made within the text in this book to 'red' and 'blue' group engines. Details of this form of classification used by the GWR relating to route restrictions is as follows — the coloured disc was displayed on the cabside.

Colour group	Axle load	GWR classes
Double Red	22½ tons	'King'
Red	Over 17 tons 12 cwt	'County', 1001, 'Saint', 3100, 3150, 'Star', 'Castle', 4200, 9300, 4700, 'Hall', 5600, 'Grange', 7200, 9400
Blue	Over 16 tons, but below 17 tons 12 cwt	1501 (certain locomotives), 1701, 1813, 'Aberdare', 2700, 2800, 'Bulldog', 4300 (except No 9300), 5100, 5700, 6100, 'Manor', 8100, 'ROD'
Yellow	Over 14 tons, but below 16 tons	2251, 'Duke', 3500, 4500, 5400, 6400, 7400, 'Dukedog'
Uncoloured	14 tons and under	517, 2301 ('Dean Goods'), 1361, 1366, 1400, 1901, 2020, 2181, 4400

The coloured disc represented the route over which the locomotive could operate. 'Red' engines were limited to main lines and lines capable of taking heavier locomotives, 'blue' engines over additional routes, 'yellow' over virtually the entire system, and 'uncoloured' engines more or less anywhere. 'Kings' had special restrictions and carried 'double red'. This form of classification was introduced by the GWR in 1920, and each coloured disc carried a letter representing the tractive effort:

Letter code	Tractive effort
Special	Over 38,000lb
E	33,001-38,000lb
D	25,001-33,000lb
C	20,501-25,000lb
B	18,501-20,500lb
A	16,500-18,500lb
Ungrouped	Below 16,500lb

Liveries

Early Cambrian locomotive livery is believed to have been green, but by 1864 black had become the standard colour. This livery continued until the Grouping.

At first, every locomotive was named, but this practice ceased after 1878, and the majority of Cambrian engines lost their names between 1885 and 1891.

From 1888, the Cambrian company crest began to appear on the leading splashers and small oval brass numberplates with a red background were introduced. In 1896, the idea of the word 'Cambrian' on one side of the locomotive, and the 'Fleur de Lys' on the other, was tried on a few splashers and was later adopted for tender and tank sides, with the crest remaining on the splashers.

Many engines later sported brass numerals instead of numberplates – the majority of these lasted until the Grouping. The number, preceded by 'No' in gold characters, shaded with black, also appeared on the buffer beams.

Below:
Sharp Stewart 0-6-0 No 898 (ex-Cambrian No 14) was one of nine 0-6-0s with 4ft 6in driving wheels, built for freight duties, which passed to the GWR. No 898, as No 14, was originally named *Broneirion* and although built by Sharp Stewart in 1875, was not purchased from the builder until March 1878, as it had originally been ordered by the Furness Railway who had not taken delivery. As can be seen in this view, No 898 is fitted with a six-wheeled tender, whereas earlier members of the class had four-wheeled versions. *E. E. Smith*

Above:
Cambrian coach No 1 was the Duke of Sutherland's private saloon (ex-Highland Railway). It was allotted the GWR number 9215 as a saloon and converted to fruit & parcels van No 1211 in January 1925, being withdrawn in December of that year. It was built in 1899 and is pictured on 27 March 1923. *British Rail*

Below:
Cambrian passenger brake van No 185 was built at Oswestry in 1900 and comprised one guard's compartment and two luggage sections. Allocated the number 117 by the GWR, this vehicle was never renumbered and was withdrawn in September 1922. *British Rail*

The locomotives were lined with a ½in band of blue grey, flanked by thin red lines. From 1904, mid-chrome replaced the grey. A new style of lining was brought in from 1915 and the word 'Cambrian' appeared in French grey block capitals on tank and tender sides, while the lining consisted of one inch French grey bands edged on either side with a ⅛in signal red line. The signal red was also used for buffer beams and wheel bosses, and buffer beam numbers were in French grey, shaded with black.

Rolling Stock

At the Grouping, the Cambrian handed over 337 standard gauge passenger vehicles, but only 221 actually carried passengers. Of these 84 were composites, about 25% were bogie vehicles, and well over half were six-wheeled stock. There were about 48 passenger brake and luggage vans (many six-wheeled) and 47 horseboxes (all four-wheeled). The remaining 21 vehicles included open and covered carriage trucks, mail, meat, milk and perishable traffic vans.

Until 1909, the carriage livery was green and white for the upper side panels and waists, and the upper side panels and waists of the ends of vehicles with observation windows were white, while the other ends, lower side panels, and all mouldings were bronze green. Underframes were black and roofs and wheel tyres were white. Lettering was gold-shaded blue and the mouldings were lined red, edged in gold and black.

In 1909, the Cambrian began to paint its carriages bronze green all over and the white upper panels were abandoned. The lining was restricted to the lower panel mouldings. The white upper panels had also been used on meat, mail and passenger brake vans, while four-wheeled milk and poultry vans, carriage trucks and horseboxes, were always painted green overall, although fully lined-out.

The next livery change came in 1915 when carriage ends were changed from green to black, roofs became grey, and lining was restricted to gold. White wheel tyres were also abandoned. Until around 1905, the ends of coaches with a guard's brake compartment were painted red.

The Company crest was etched on frosted glass toilet windows and shown on panels of all coaches from 1886. Class designations were written in full on coach doors and placed between the waist mouldings with the Company's name and the coach number. Others were marked 'LUGGAGE' or 'GUARD'.

From 1900, the words 'CAMBRIAN RAILWAYS' were moved from the waistline to between the mouldings above the carriage windows. The 'Fleur de Lys' emblem was shown on the lower sides of all bogie coaches and on a few special four-wheelers and six-wheelers.

After the Grouping, the GWR soon replaced the older Cambrian carriages with relatively new corridor coaches. Corridor stock was a new luxury on Brecon trains, but on the branch lines the old Cambrian coaches were replaced by equally ancient GWR compartment stock – much of it four-wheeled. The GWR did not introduce modern compartment stock on to the Cambrian until the mid-1930s. By nationalisation in 1948, there was a good proportion of Hawksworth corridor stock on the system, although some Cambrian bogie stock was still to be found.

Cambrian special vans were soon replaced by GWR vehicles of a similar type. The postal sorting van was replaced in 1933 by a new GWR Travelling Post Office but, in later years, mail was handled in ordinary luggage vans.

The freight stock at the Grouping comprised 2,278 standard gauge goods and service wagons, including 1,517 open wagons (109 for loco coal and 62 for ballast), 372 timber, rail and flat trucks, 191 cattle wagons, and 110 covered goods wagons. There were also 10 travelling gas holders, 38 goods brake vans, and eight other wagons (specials for unusual loads, including boiler trucks and gunpowder vans – all metal, and similar to those on the GWR). The balance was made up of 14 cranes, and 18 service vehicles, like mess and tool vans.

Goods livery was medium/light grey with the ironwork/strapping in black, and solebars in grey. Lettering was white, shaded in black. Brake vans had vermilion ends until around 1905. Roofs were white on vans as were the lower panels on cattle wagons. The Railway Clearing House insisted, as CR was used by the Caledonian Railway, that a different identification was used. The style chosen was 'CAM RYS' with a central 'Fleur de Lys' emblem, but vans and cattle trucks continued to display 'Cambrian Railways' in full (in small letters). From 1915, wagons were relettered 'CAMBRIAN' in large capitals.

The 'Cambrian Coast Express'

The inaugural run of the 'Cambrian Coast Express' was in 1927, although in 1921, an express operated from Paddington to Aberystwyth, Barmouth and Pwllheli which included a restaurant car. This train was not named but was the precursor of the 'Cambrian Coast Express'.

By 1923, this train, which was now leaving Paddington at 10.20am, became a Birkenhead train, and after a non-stop run to Birmingham (Snow Hill), an Aberystwyth portion was detached.

After various departure time changes over the years to 1927, when the train was officially named, it became the 10.10am off Paddington during the summer months and operated from Mondays to Fridays only. In 1939, the 'Cambrian Coast Express' became a Saturdays Only working.

Generally, the train would be hauled from Paddington by a GWR 'Castle' which was taken off at Wolverhampton (Low Level). Here, because of the light track and bridges over the Cambrian, one or two 'Duke' class locomotives would take over. In latter days this, of course, would be a GWR 4-6-0 'Manor'.

The reason that the locomotives were changed at Wolverhampton was due to the fact that the train ran non stop from there to Welshpool, by using the Abbey Foregate curve at Shrewsbury. After Welshpool, the 'Cambrian Coast Express' stopped at Machynlleth, Dovey Junction and Borth, arriving at Aberystwyth at 3.55pm.

The up 'Cambrian Coast Express' left Aberystwyth at 10am and also stopped at Newtown before arriving at Wolverhampton for its change of motive power at 1.28pm. After its 19min run to Birmingham Snow Hill, the train departed for Paddington at 2pm, taking 2hr to reach its London destination – the up journey taking 25min longer than the down run over the 234¾-mile route from Aberystwyth.

After the war, the 'Cambrian Coast Express' was reinstated – again Saturdays Only – but the title was not used. It still departed from Paddington at 10.10am and became the only non stop run of the day to Birmingham Snow Hill. The same intermediate stations were called at on its way to Aberystwyth, but Newtown was included by both up and down trains.

In 1951, the name 'Cambrian Coast Express' was restored and the train became a daily working, stopping at Banbury en route to Birmingham. Also, this train no longer ran non stop from Wolverhampton to Welshpool, but ran into Shrewsbury where it reversed. Aberystwyth was reached in just under 6hr from Paddington. At Machynlleth, through coaches for Barmouth and Pwllheli were detached.

The restaurant car offered light refreshments and lunch at reasonable prices and, as an example, in 1960, coffee and biscuits cost 1/4d (7p), luncheon was 11/- (55p), and afternoon tea 3/6d (17½p). You could even get a glass of whisky (called GWR Special) for 2/6d (12½p), and beer was 1/7d (8p) per bottle – those were the days!

Below:
The Aberystwyth portion of the 'Cambrian Coast Express' prepares to depart from Dovey Junction on 25 September 1961. GWR 'Manor' No 7803 *Barcote Manor* displays the official headboard. *R. F. Roberts*

Bottom:
Carrying the alternative 'Cambrian Coast Express' headboard, which excluded the crest, Standard 2-6-0 No 78006, in green livery, leaves the Mawddach estuary viaduct and approaches Barmouth, on 3 July 1957, with its three-coach train.
T. E. Rounthwaite

The end for the 'Cambrian Coast Express' came in 1967 as, from March, trains from Paddington were diverted to Birmingham New Street, and Aberystwyth was served by Euston to Shrewsbury trains, via Wolverhampton – Saturdays Only during summer months, with no winter through service.

The final steam working on the Cambrian main line was on Saturday, 4 March 1967. The last 'Cambrian Coast Express' from Aberystwyth to Paddington was hauled, as far as Shrewsbury, by BR Standard Class 4 4-6-0 No 75033. Harry Rees, the Aberystwyth Stationmaster, turned out in full Great Western Stationmaster's uniform to see the train off.

The last down 'Cambrian Coast Express' was hauled by No 75021, and this locomotive was also used on the last up Mail.

Accidents and Mishaps

The Cambrian Railways were a little slow to provide safety devices on locomotives and rolling stock, and the first locomotives built with vacuum brakes were the two 'Beaconsfield' 4-4-0s, built in 1886. This was followed by a further 14 passenger locomotives together with carriage stock to be so fitted, enabling main-line passenger trains to be automatically braked. Completion of automatic braking on Cambrian stock was in the mid-1890s.

Another safety device to be introduced was the single line train staff, to prevent two trains using the same section of single-line track and, following the introduction of the telegraph on the main line to Aberystwyth, in 1865, a better system, the staff and ticket, was introduced. This gave stationmasters the power to issue written authority for drivers to proceed and retain the staff for the next train.

It was 1890 before the main line was protected, after the introduction of Tyer's electric tablet, in 1888. Fortunately slow-speed running over the Cambrian prevented many minor mishaps turning into major disasters, and not until 1897 was a passenger killed on the Cambrian.

However, in November 1861, Abermule witnessed an incident where a driver tried to make up time lost having a quick drink by travelling too fast over a stretch of track only completed four months previous. The locomotive overturned and sadly both driver and fireman were killed.

Abermule was also the scene of the Cambrian's worst disaster. On 26 January 1921, due to an extraordinary and tragic misunderstanding, a slow down train was given the go ahead to proceed towards Newtown, meeting, head on, an express from Aberystwyth, a mile from Abermule, on a curve. The death toll, the largest in Cambrian history, was 17.

The fireman of locomotive No 82 actually took back the Montgomery to Abermule tablet, the section he had just passed over, from a young signalman. The inevitable, of course, happened. No 82, whose crew were killed in the incident, met No 95 on the express from Aberystwyth. The Company Chairman, Lord Herbert Vane Tempest, along with 16 other passengers, was also killed.

Two sad incidents have occurred at Friog over the years, the first in 1883 and the next in 1933. On both occasions the driver and fireman were killed. On 1 January 1883, the exhaust of a steam locomotive set off a landslide after a day of heavy storms.

Along this stretch of coastal track the speed limit was just 4mph, and the 5.15pm Machynlleth - Barmouth train had just shut off steam at the top of the 1 in 66 bank from Llwyngwril and, in the darkness of that night, 120 tons of soil and rock, together with a 30ft section of the wall beside the road above the railway, came cascading down, knocking the locomotive, No 29 *Pegasus*, over the cliff-side to the sea below. Miraculously, although the empty first coach followed the locomotive over the edge, the strength of the couplings between the second and third coach, which remained on the track, prevented the other coaches plunging down the cliff.

Right:
The Friog avalanche shelter, constructed after the 1933 accident, with Standard 2-6-4T No 80101 approaching with a northbound passenger train. *C. L. Caddy*

Left:
A slight mishap at Barmouth Junction in February 1961 resulted in LMS Fowler 2-6-2T No 40110 leaving the track. At the time, this locomotive was allocated to Chester LNWR — 6A.
Author's Collection

Below:
Another mishap — this time at Machynlleth where it was found that the wagon shunted by Standard 2-6-2T No 82021 did not 'fit' in the fitting shop!
Author's Collection

In the accident at Friog on 4 March 1933, the 6.10 am mail train from Machynlleth to Pwllheli hit debris which had fallen from the road above. Only the locomotive plunged over the cliff to the rocks below. Passenger carriages remained on the trackbed but the locomotive, No 874, had to be cut up on the beach and became just scrap metal.

An avalanche shelter was constructed after the 1933 Friog accident, part of the stabilisation work which continued for three years. The road above the railway was rerouted at the same time.

On 18 January 1921, 'Jones Goods' locomotive No 54 was derailed in Talerddig Cutting following a landslide which stopped through services for two days.

In addition to the major disasters, the Cambrian, of course, had its share of minor incidents.

Severe Weather Conditions

Storms, gale-force winds and heavy snowfall were conditions which regularly caused problems over the Cambrian system during steam days.

In September 1922, flooding at Aberystwyth was so severe that the original Vale of Rheidol station was 3ft under water. The main line station was also flooded and, on shed, the water level, caused by the rising of the River Rheidol, was up to firebox level.

Five years later, in 1927, huge waves came up the Dovey estuary, and pounded the track where No 3270 *Earl of Devon* was heading after negotiating deep flood water at Dovey Junction. At Glandyfi, the train was stopped, and passengers spent the night in the small station house.

The cruel winters of 1947 and 1962 brought deep snow and, in 1947, at Talerddig Summit, the snow was up to the tops of the telegraph poles. Sadly, when the thaw came, the swollen rivers brought down the mountains thousands of dead sheep that had perished in the snow.

The 1962/63 winter was even more severe. Collett 0-6-0s carried permanently-attached snowploughs. Some lines were frozen solid – even sand dunes were rock-hard. Water mains froze up and it was amazing that the railway was able to continue operating.

Above left:
A typical freeze-up at Talerddig cutting — at least the line was clear on this occasion! *Author's Collection*

Left:
Severe flooding at Dovey Junction as 'Manor' No 7821 *Ditcheat Manor* prepares to plough through the deep water. *C. L. Caddy*

Barmouth Bridge

When the line to Barmouth Junction from the south was completed in July 1865, it was thought that the only way to reach Barmouth across the Mawddach estuary was by the long way round – along the branch to Dolgelley. But until 1869 the line only ran as far as Penmaenpool. In June of that year, the line was continued to Dolgelley, using a platform some distance from the town. Dolgelley station proper was not used by trains from Barmouth Junction until the GWR completed its line from Ruabon.

There were grave doubts in the minds of many regarding the practicability of crossing the Mawddach, with its mass of sand and its river current, into Barmouth, and it required tremendous scientific skill and courage to take on this task.

However, before long, what at first appeared to be impossible, became a reality. Over 500 timber piles in 113 spans were driven into the sand, making up the greater portion of the first 800yd. The navigable channel at the Barmouth end of the bridge was traversed by an iron construction of seven fixed and one opening span. This opening span was of the drawbridge type, and was carried on wheels, when lifted, by means of large screws at one end, and could be drawn back over the adjoining span.

This process of opening the bridge to river traffic was long and arduous and, in 1899, when the viaduct had become too weak for the ever increasing traffic, it was replaced by a modern steel structure of four spans, one of which was sprung, and revolved on the centre pier, giving two clear openings. The piers comprised 8ft diameter columns which were sunk into the sand and into solid rock, reached at around 90ft below the high water mark. The columns were of steel and filled with concrete. The contractor was the Cleveland Bridge and Engineering Co Ltd of Darlington.

From 1906 to 1909, the timber section of the viaduct was also renewed, again in timber, by the firm of Abraham Williams of Aberdovey, who had been involved in the construction of many of the old wooden bridges on the coastline. During the entire renewal work, there was never a moment's stoppage in traffic – some achievement!

Even the original viaduct was, for its day, a tremendous engineering achievement, and its footpath beside the railway provided Barmouth with a promenade with a wonderful panoramic view of sea, river and mountains, unrivalled anywhere in Britain.

Prior to completion, a horse-drawn carriage used the bridge for a time until the bridge opened to rail traffic in 1867. Unfortunately, in 1980 a tiny sea animal, the teredo shipworm, discovered by divers when carrying out a routine check of the structure, had caused an estimated £2½ million of damage to the viaduct. Over the years, splicing repairs have been carried out when necessary to the pitch pine piles which support the timber spans.

Deterioration mainly occurs between tide levels. By 1980 a total of 69 piles, on inspection, had been affected by the teredo shipworm. It apparently has two small shells in its head which it uses for boring into the timber. Its larvae had made pin-size holes to get inside the timber piles, where it lived. These holes increased to ½in diameter and reached several feet in length.

The teredo shipworm can live only in continuous salty water and it was found that new piles were attacked within five weeks. Emergency repairs were carried out and a 10mph speed restriction was imposed – with no locomotive-hauled trains being allowed over the viaduct.

Eventually, full repairs were carried out, involving the grouting of the holes made by the teredo shipworm, with resin, encasing the wooden piles in glass-reinforced plastic, and building stone embankments. Locomotive-hauled traffic can now traverse the bridge with confidence.

Below:
Barmouth Bridge which passes over the Mawddach estuary. This view, photographed on 18 September 1977, shows the movable section in the 'shipping access' position. *C. L. Caddy*

Bibliography

Rex Christianson & R. W. Miller; *The Cambrian Railways* (Vols 1&2); David & Charles

W. J. K. Davies; *Vale of Rheidol Light Railway*; Ian Allan

R. W. Kidner; *The Cambrian Railways*; The Oakwood Press

C. P. Gasquoine; *The Story of the Cambrian*; Woodhall, Minshall, Thomas & Co Ltd

T. P. Dalton; *Cambrian Companionship*; Oxford Publishing Co

Locomotives of the Great Western Railway; RCTS

Railway Magazine (various issues)

E. Lyons; *An Historical Survey of Great Western Engine Sheds 1947*; Oxford Publishing Co

R. H. Clark; *An Historical Survey of Great Western Stations*; Oxford Publishing Co

W. G. Rear & M. F. Williams; *The Cambrian Coast Railway*; Foxline Publishing

1:

Whitchurch to Oswestry

Of the original four railway companies which were amalgamated into the Cambrian Railways Co on 25 July 1864, the Oswestry, Ellesmere & Whitchurch Railway was the last to open – the Ellesmere to Whitchurch section to freight on 20 April 1863 and to passengers on 4 May 1863, and the Oswestry to Ellesmere section on 27 July 1864.

In September 1858, the LNWR completed its single line from Crewe to Whitchurch (which was doubled in 1862), and it was stated at an inaugural meeting that the local company's new line from Oswestry to Whitchurch would not only serve the local area but would link with routes to the north, via Crewe and Manchester. A proposed branch from Whittington to Gobowen was abandoned by the proposers of the OE&WR after hearing of the GWR's intention to include Oswestry on its main line to Shrewsbury from Chester, in the form of a loop. The OE&WR was built by Thomas Savin and John Ward, and covered a distance of 18 miles.

Inclines over the route were quite severe in certain areas, the steepest being each side of Frankton, where there were two stretches of 1 in 80 gradient, each a mile in length, the centre of which was a summit half a mile on the Oswestry side of Frankton. There was another 1 in 80 incline between Fenn's Bank and Whitchurch, about half a mile in length.

The station at Whitchurch was, in fact, an LNWR station – the Cambrian line joined the LNWR Shrewsbury to Chester/Crewe line at Cambrian Junction, 8 chains south of Whitchurch. Therefore, no views of Whitchurch station will be found in this Cambrian album. The first station on the line was at Fenn's Bank, 2 miles 72 chains from the junction. Whixall Moss, a large bog, lay between Fenn's Bank and Bettisfield. The draining of this bog was considered, but this did not come to fruition and it remained a picturesque expanse of heather and silver birches, rich in peat. The surveyors, during their soundings regarding the possibility of the bog taking the weight of passing trains, lost their 35ft rod in the swamp, but this did not deter them from working out a way to avoid the railway meeting a similar fate. The moss was, on average, about 12ft deep and, at the deepest point, 17ft, and beneath this there was sand. It was near the canal which crossed the swamp where the surveyors lost their rod, where it was deeper.

Their solution was to build the line on a raft of brushwood, 3ft thick, and between 8ft and 12 ft long, supported on a framework of larch poles extending the full width of the rails. These poles were then interwoven with hazel branches and brushwood and, upon all this, the sleepers and rails were laid, ballasted with sand. This section of track covered a distance of almost 3 miles. By spring 1863, the railway was ready to take traffic over the 11 miles from Whitchurch to Ellesmere.

The first passengers to travel the 11-mile route were Capt Jebb and his Company of Rifles, on 20 April 1863, together with a few friends. An official trial run soon followed with *Montgomery* and *Hero* double-leading the train and, on the first Monday in May, the first regular service commenced from Ellesmere to Whitchurch. By 1923, nine trains a day worked passenger services between these two locations, two in each direction running non stop, and one working departed from Liverpool at 12 noon, calling at Whitchurch at 1.50pm, and then ran non stop to Oswestry where it arrived at 2.28pm. There was one Sunday train in each direction.

Welshampton, west of Bettisfield, was the scene of an accident in June 1897. A Sunday School excursion returning from Barmouth to Royston in Lancashire, double-headed, left the rails, several coaches overturning and others telescoping. Sadly, nine children lost their lives in the incident and two died in hospital. Many others were injured. The tragic accident was attributed to the 'rough running' of a four-wheeled L&Y brake van, situated at the front of the train.

Prior to this incident, in 1887, the down morning mail train ran off the line at Ellesmere, all due to the fact that the porter was not in position at the points in time to operate them properly. Interlocking of points was not made compulsory until an Act in 1889. The porter was consequently dismissed, and the Ellesmere stationmaster, who signed a petition requesting his reinstatement by the management, was also temporarily suspended and removed from Ellesmere, as the management felt that it was the wrong attitude for a stationmaster to take over a matter of service discipline. Mr Hood, the stationmaster, was summoned to Crewe for an interview in front of the Company Chairman and Board members.

As reinstatement was refused, the incident blew up into a major talking point in the industrial world. Protest meetings were held in Wrexham and Newtown, condemning 'excessive working hours', the reason claimed as contributing to the incident. The Amalgamated Society of Railway Servants of the United Kingdom claimed that 'little railways were a gigantic mistake', and the Labour press shouted 'Cambrian tyranny' and 'victimisation'.

At Ellesmere, which was eventually to be joined by a Cambrian line from Wrexham Central in 1895, work commenced on the 7½-mile section to Oswestry on 4 September 1862, but at first, Whittington was the only station between the two locations. Frankton station, which served two border villages (English Frankton and Welsh Frankton) was constructed in 1866, and even then a permanent station was not built there until the early 1880s at the same time as one was built at Fenn's Bank. In December 1868, the platform at Whittington was moved from the south side of the line to the north side (the budget for the operation was a mere £6).

Just west of Whittington, the GWR Shrewsbury to Chester line passed beneath the Cambrian, and the line ran into

Oswestry. From 1922, the Cambrian station at Whittington was called Whittington High Level.

As the Cambrian line entered Oswestry it joined the Great Western line from Gobowen. The Gobowen shuttle from Oswestry was regularly photographed by enthusiasts in steam days, and saw a vast variety of motive power, and auto-sets with two trailers ran over the line, often with an extra trailer attached and sometimes with two sets running together as detailed in the summer 1946 'Working of Coaches' schedule. Great Western 'Manors' were introduced in 1938, and on 8 November permission was granted to operate these locomotives over the Oswestry to Gobowen line, and also over the Whitchurch to Oswestry section, the maximum overall speed not to exceed 40mph between Fenn's Bank and Bettisfield. However, at that time, they were not allowed in the down loop or down refuge siding at Fenn's Bank, or in the loading wharf at Bettisfield. The notice, in fact, gave permission for any 'blue' group locomotives (2-8-0s excepted) to traverse this line, bringing in GWR 2-6-0 Moguls (except 9300 class).

Regarding locomotive workings in the area, a notice from the Divisional Engineer's office at Oswestry, on 9 July 1936, gave permission for GWR 4-6-0 'Granges' ('red' group) to work from Oswestry to Gobowen and, in fact, on 19 May 1939, permission was granted for any class of locomotive to work the line except 'Kings' and 4700 class, with certain yards and sidings being out of bounds. However, in the 1936 notice, it was pointed out that the Gobowen branch was the only portion of the Central Wales Division over which 'red' group engines would be permitted. In January 1945, coaching stock 63ft 6in long 9ft 7in wide was allowed into Oswestry and beyond from both the Whitchurch and Gobowen direction. The branch from Gobowen, at first, went into a separate GWR station at Oswestry on the west side of the Cambrian station, and the line was joined to the Cambrian line by a spur, 14 chains in length (Cambrian-owned). Oswestry Works was situated on the left of the Cambrian line on entering the station from Whitchurch, and was directly opposite the GWR station. The Works was built at Oswestry despite strong protests by the people of Welshpool who claimed that their town was more central, but as land had already been bought at Oswestry, the main locomotive, carriage and wagon works was constructed there at a cost of £28,000. A small repair shop was maintained at Welshpool by Savin.

The Works came into operation in January 1866, but only on a limited basis, and after Savin's bankruptcy, the Cambrian completed the job, by August 1866. The Welshpool workshop was then closed. An additional wagon repair shop was opened at Aberdovey in 1880 as Oswestry was so far from the coast line.

At Oswestry, the entire locomotive, carriage and wagon shops were housed in a single building, 812ft long × 210ft wide, covering an area of 170,520 sq ft, using 4,145,952 bricks and 28,483ft of timber.

The Great Western station at Oswestry closed on 7 July 1924, GWR traffic being diverted to the Cambrian station, and the Gobowen branch closed to passenger traffic on 7 November 1966. The Oswestry to Whitchurch line had previously closed to passengers on 23 November 1964.

Below:
With a Standard 2-6-4T in charge, the 4.30pm Whitchurch to Welshpool train is seen at Bettisfield on 28 August 1964, less than five months before the line closed to passengers.
Andrew Muckley

Above:
On a frosty morning in January 1965, the first Whitchurch to Oswestry working of the day heads westwards from Bettisfield, hauled by an Ivatt Mogul. *J. I. C. Boyd*

Below:
Although the GWR 4300 class Moguls were introduced in 1911, it was 1939 before any locomotives of this class were allocated to Central Wales. Here, No 6375 is seen approaching Ellesmere, from Whitchurch, with a three-coach train. The goods shed and yard are seen in the background. *K. Smith*

Oswestry

Ex-Cambrian Stock

Above:
On 28 August 1952, a Belpaire 15 class 0-6-0, No 887 (ex-Cambrian No 89) stands in steam outside Oswestry shed, proudly displaying its 'lion on wheel' emblem on the tender. The locomotive was withdrawn three months later. *Brian Morrison*

Below:
An ex-Cambrian 0-6-0, No 45, photographed in the 1930s when carrying the GWR number 900, stands at Oswestry shed. This locomotive was originally named *Rhiewport* and was built in 1864 and withdrawn in September 1945.
H. N. James

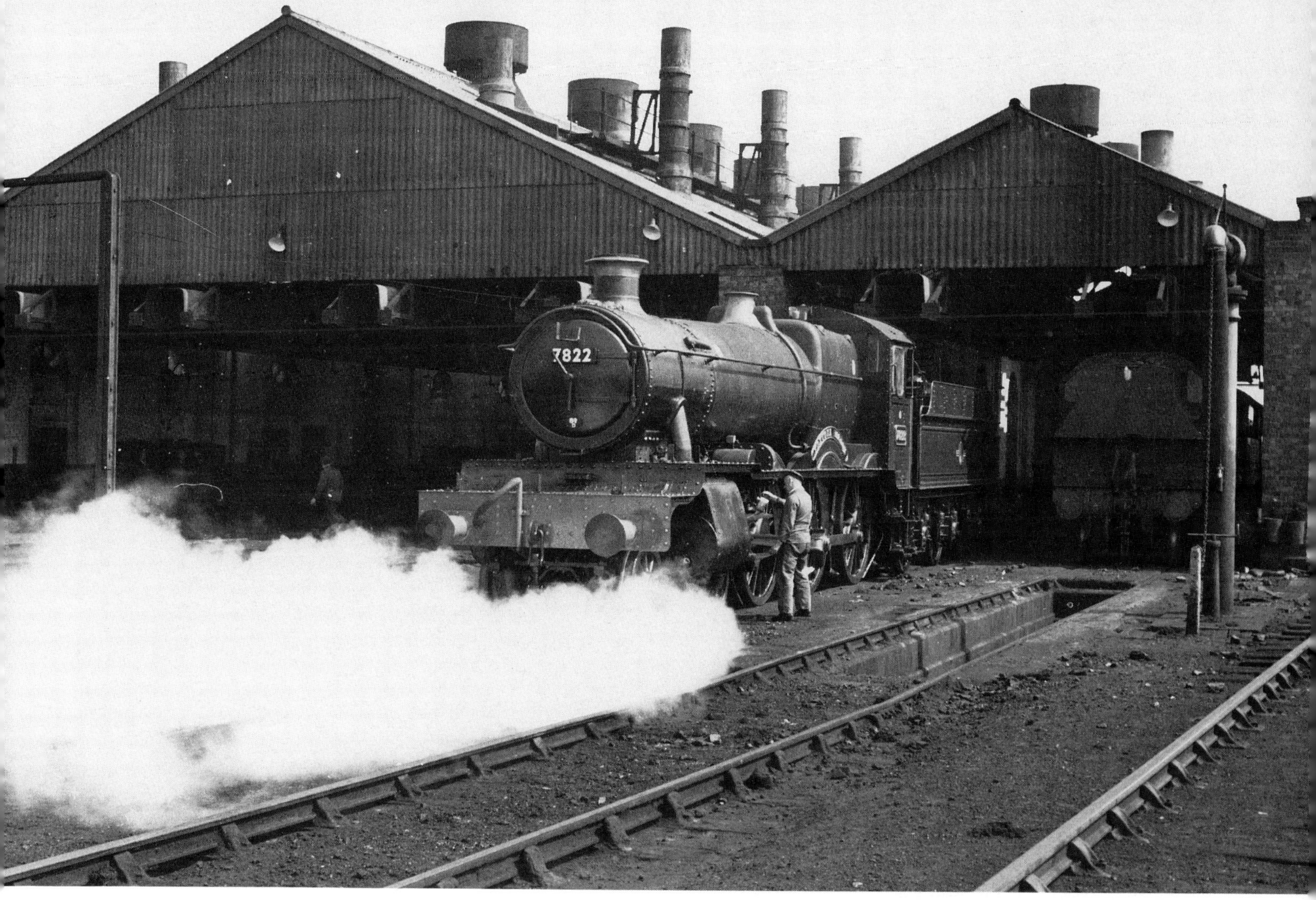

Foxcote Manor on shed

Above and below:

Two views of GWR 'Manor' No 7822 *Foxcote Manor* at Oswestry shed. The views show clearly the structure of the shed building and the turntable in operation (manually). This locomotive is now one of the nine 'Manors' preserved, and can be found at Llangollen, and is operational. Oswestry's shed opened in 1860 and closed in January 1965 and the main building measured 200ft × 67ft. The roof seen in these views was added in 1939, together with other improvements.
K. Smith

Oswestry Tanks

Above:
Great Western 1400 class 0-4-2Ts were allocated to Oswestry for auto-train duties to Gobowen and for Wrexham to Ellesmere trains. There was also an Oswestry to Tinkers Green shuttle service in the 1940s. No 1458 is pictured standing over the pits at Oswestry. *N. E. Stead*

Below:
In 1866, Sharp Stewart built three small 2-4-0T engines which were originally used on the Dolgelley and Llanfyllin branches, one of which was withdrawn in August 1929 (GWR No 1192). However, Nos 1196 and 1197 (ex-Cambrian Nos 58 and 59) lasted until April 1948. No 1196, seen here in Oswestry's shed yard, in the 1930s, worked its days out with No 1197 on the Tanat Valley line to Llangynog. Both engines covered a million miles in their lifetime. *H. N. James*

Above right:
The 7400 class pannier tanks were also fitted for push-pull auto-train working and, on 22 July 1954, No 7405 was seen taking on water at Oswestry shed before returning to duty. Around this time, three members of this class were allocated to Oswestry. *Brian Morrison*

Right:
On 22 July 1954, GWR Prairie tank No 4546 simmered off duty at Oswestry. This was Oswestry's sole member of the class at the time, although Machynlleth had a larger allocation.
Brian Morrison

7405

4546
4546

Oswestry Works

Right:
Mawddwy Railway engine No 824 (formerly Cambrian No 30) was built in 1865 and lasted in service until September 1940. It was built by Manning Wardle and was an 0-6-0ST. Originally named *Mawddwy*, it also worked the Van branch. Its last days were spent working the Tanat Valley line. It is seen inside Oswestry Works during the 1930s — note the most unusual cab. *H. N. James*

Left:
The two Welshpool & Llanfair narrow gauge locomotives, Nos 822 and 823, were moved to Oswestry Works when the line was originally closed to freight traffic by British Railways in November 1956. Prior to purchase by the preservation group, these locomotives could be found side by side, stored in the Works. On 14 August 1958, they were seen there minus their nameplates.
Brian Morrison

Right:
Ex-Glasgow & South Western Railway 0-6-0T No 9 was built by North British in 1917, renumbered to 324 in 1919, and eventually carried the LMS number 13679. In 1934 it was sold for work at Llay Main collieries in Denbighshire. it is pictured in the Cambrian workshops at Oswestry prior to being preserved in 1965, and is at present displayed in pristine condition in the Glasgow Museum of Transport. This outside-cylinder 0-6-0T is seen fitted with a spark arrester. *E. N. Kneale*

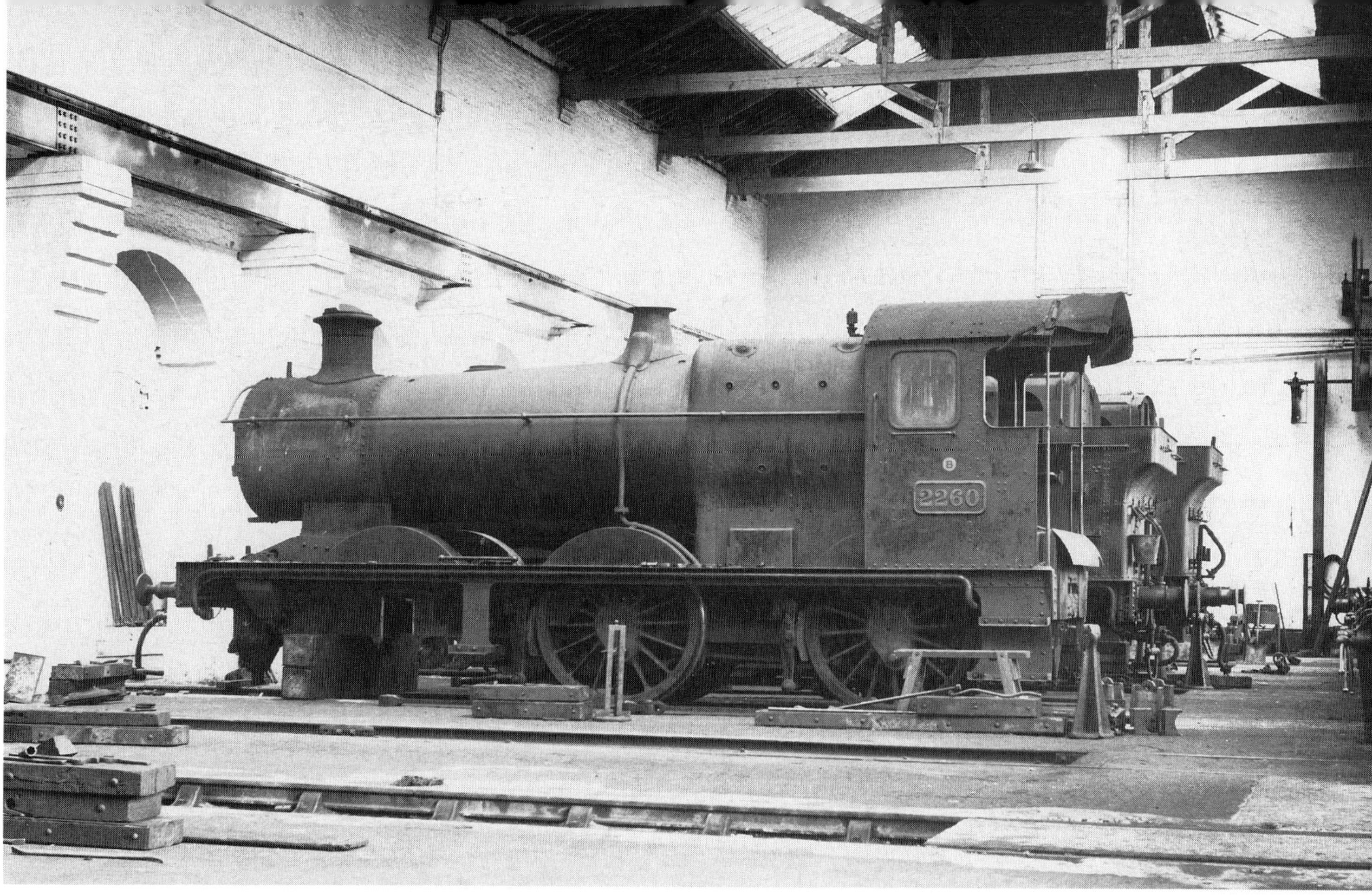

Above:
Minus its front driving wheels, GWR 2251 class 0-6-0 No 2260, clearly showing its yellow circle and 'B' power class designation, awaits attention in Oswestry's Cambrian workshops on 22 July 1954. This Machynlleth-allocated engine was one of 12 Collett 0-6-0s on the Central Wales Division around this time. *Brian Morrison*

Below:
A larger than usual visitor to the Cambrian workshops at Oswestry was this LMS 'Jubilee' No 45572 *Eire*, seen awaiting repair. *E. N. Kneale*

Freight Operations

Above:
Great Western 'Manor' No 7819 *Hinton Manor* carries out shunting duties in Oswestry's freight yard on 23 June 1956. This locomotive is now preserved on the Severn Valley Railway and is the only member of the class currently passed for main-line running. *R. E. Vincent*

Above right:
With a mixed freight, Ivatt 2-6-0 No 46521 hauls its train from Oswestry station on 26 July 1960. An auto-train can be seen in the bay platform. This engine was one of the first two members of the class to arrive at Oswestry early in 1953 (together with No 46503), mainly for working Whitchurch and Ellesmere trains. *S. D. Wainwright*

Right:
The 1600 class 0-6-0PTs were Hawksworth engines, built for light branch-line work and shunting duties. They were not introduced until BR days, in 1949, and in their early years only two could be found on the Central Wales Division, Nos 1603/04. The class of 70 locomotives was not completed until the mid-1950s. No 1638 is seen on shunting duties at Oswestry station. *E. N. Kneale*

6521
4521

Above:
The penultimate member of the 1600 0-6-0PT class, No 1668, pauses during shunting duties at Oswestry on 8 August 1962. The life of these pannier tanks was comparatively short due to dieselisation. By the early 1960s, some had already been withdrawn as they had become redundant. In 1935 Oswestry had still used shunting horses, along with 27 other GWR stations. *S. D. Wainwright*

Lady Margaret

Above right:
Ex-Liskeard & Looe Railway 2-4-0T No 1308 *Lady Margaret* arrived at Oswestry in 1922, and remained there until withdrawal in May 1948, apart from one short spell from 1927 to 1931 when it returned to the West Country. Used regularly on the Tanat Valley line, No 1308 is pictured at Oswestry during the 1930s. *H. N. James*

Right:
The winter of 1962/63 was one of the worst for many years. On New Year's Day, 1963, a Whitchurch train awaits departure from Oswestry, hauled by GWR 'Manor' No 7807 *Compton Manor*. *J. I. C. Boyd*

7807

Oswestry to Gobowen Shuttle

Right:
Various types of locomotive were used on the Oswestry to Gobowen trains and, in this view, Belpaire 15 class 0-6-0 No 895 (ex-Cambrian No 101), built in 1908, is seen at Oswestry having arrived with the 1.25pm train from Gobowen, on 22 July 1954. Built originally for freight work, and until 1908 numbered 42, this engine was withdrawn in October 1954.
Brian Morrison

Below:
Croes Newydd-allocated 1400 class 0-4-2T No 1473 prepares to leave Oswestry, on 28 August 1952, with the 10.35am service to Gobowen. These trains ceased operation on 7 November 1966, closing Oswestry station completely to passengers. Whitchurch to Oswestry trains had already ceased on 18 January 1965.
Brian Morrison

Above:
With an auto-train for Gobowen, 5400 class 0-6-0PT No 5422 leaves the bay platform at Oswestry station with the 3.45pm train, on 31 July 1959. At this time, two engines of this small class were allocated to Oswestry — Nos 5400 and 5422.
S. D. Wainwright

Below:
An Ivatt 'Mickey Mouse' 2-6-0, No 46516, approaches Oswestry station off the branch with the 3.10pm train from Gobowen, on 26 July 1960. On 22 April 1940, GWR diesel railcars were given permission to operate this branch (except cars Nos 17 and 18) and, in fact, on 21 November 1946, the experimental gas turbine locomotive, No 18100, was also given clearance, as this was a 'red' route. *S. D. Wainwright*

'Dukedogs'

Bottom:
At the head of the 2.05pm Whitchurch to Welshpool train, on 22 July 1954, 'Dukedog' No 9005 takes on water at Oswestry station. Renumbering from the 3200 series to the 9000 series took place in 1946. *Brian Morrison*

Below:
'Dukedog' No 9002 hauls the 11.15am Whitchurch train out of Oswestry station on 28 August 1952. By May 1953, the only 'Dukedogs' not allocated to the Central Wales Division were Nos 9008/10/16/20 (in store), 9015 (Didcot, 81E), 9018/28 (Oxford, 81F) and 9011/23 (Swindon, 82C). No 9002 was withdrawn in May 1954. *Brian Morrison*

Right, top and bottom:
On 28 August 1952, the 10.50am train for Pwllheli was hauled into Oswestry by 2301 class 'Dean Goods' No 2411. From here, 4-6-0 'Manor' No 7819 *Hinton Manor* took over and can be seen alongside. In the other view, the 'Manor' has taken over and awaits departure. Around this time six 'Manors' were allocated to Oswestry. *Brian Morrison*

2411

7819

2:

Ellesmere to Wrexham Central

The Wrexham & Ellesmere Railway opened to traffic on 2 November 1895, 31 years after the formation of the Cambrian Railways Co. Although an independent company, the W&ER was worked by the Cambrian from the outset. An original suggestion that a line from Wrexham Central should be built to Bettisfield by the Wrexham, Mold & Connah's Quay Railway was not taken up and the line was built to Ellesmere.

The railways of Wrexham formed a complex set-up involving the Great Western, Great Central and Cambrian railways, the Cambrian using the Great Central station – a short distance away was the GWR station and Wrexham Exchange, owned by the GCR. The depot at Croes Newydd was GWR-owned.

There were only three intermediate stations on the line between Wrexham and Ellesmere – Marchweil, Bangor-on-Dee and Overton-on-Dee, but there were halts at Hightown, Sesswick, Pickhill, Cloy, Trench and Elson. Passing loops were at the three stations – Marchwiel (771ft), Bangor-on-Dee (804ft) and Overton-on-Dee (675ft). In 1923, nine trains in each direction worked the line, six of which were rail motor car duties – one class only. The journey time was around 32min, trains also calling at Trench, Sesswick and Hightown halts – the line was 12¾ miles long. The 1904 working timetable shows three freight workings per week in each direction – from Ellesmere, one Mondays Only general freight, one goods and cattle on Wednesdays Only, and another general freight working (not Wednesdays). In the opposite direction there were two daily freights, and a cattle train to Whitchurch, only on Mondays – freight trains making up a third of the line's total workings.

In December 1894, Nasmyth Wilson submitted a drawing for a 4-4-0 side tank with 3ft 1½in bogies and 5ft 3in coupled drivers. This was generally accepted but the wheel arrangement was changed to 0-4-4. These locomotives first worked the Wrexham to Ellesmere branch, three being

delivered in 1895 and the final three in 1899. They were numbered 3/5/7/8/9/23 (GWR numbers 10/11/15/19/20/21) and all were withdrawn by October 1932. GWR locomotives 11, 15 and 21 never carried their allotted numbers. These engines also carried out duties on the Llanfyllin branch, and later some were found at Penmaenpool, working through trains to Barmouth from the Great Western at Dolgelley. The last locomotives to work passenger trains over the route were the GWR 1400 class 0-4-2Ts, introduced on to the Great Western in 1932, and the last passenger train ran on 8 September 1962. The old Great Central line from Bidston, near Birkenhead, is still open to passenger traffic, and a short stretch of ex-Cambrian line to Abenbury sidings from Wrexham is still open to freight.

At the Ellesmere end, the Wrexham branch joined the Ellesmere to Oswestry line just 10 chains from the station, this line not featuring in the many proposals of lines to converge on Ellesmere in 1861. The OE&WR's branch to Ffrith, north of Wrexham, was never built. This would have continued south of Ellesmere to Wem, on the LNWR, and a GWR line from Whitchurch, continuing to Rednal on the GWR's Shrewsbury to Wrexham line was to be built. Additionally there was the Wrexham, Mold & Connah's Quay Railway's proposal to build a line from Wrexham to Bettisfield.

At Wrexham Central, the combination of both GWR and GCR/LNER locomotives added particular interest to this location. In addition to the GWR 0-4-2 tanks and pannier tanks, LNER C13/C15 4-4-2Ts and N5 0-6-2Ts, allocated to Wrexham (6E), could be found working in from the north in the early 1950s.

Left:
The 1400 class 0-4-2Ts were regularly employed on Wrexham to Ellesmere trains until closure of the line to passengers from 10 September 1962. The locomotives used were allocated to Oswestry and No 1458 is seen with its single-coach train at Ellesmere. *K. Smith*

Above:
Just two months prior to termination of the line's passenger services, 0-4-2T No 1438 stops at Bangor-on-Dee, on 27 July 1962, where parcels are unloaded from its train. Note the overgrown state of the platform. *G. Coltas*

Below:
A typical push-pull working on the Ellesmere to Wrexham branch pauses at Bangor-on-Dee in May 1962. One of the regulars, 0-4-2T No 1432, is in command. *J. I. C. Boyd*

Bottom:
After closure of the Wrexham to Ellesmere branch, the line was truncated, but freight trains still ran daily from the Wrexham end. On 19 August 1966, a 0-6-0PT tank, No 3709, drifts down to Marchwiel with a train of just four wagons and a brake van. *Les Nixon*

Wrexham Central

Above:
In January 1952, an Ellesmere auto-train stands at Wrexham Central, headed by GWR 0-4-2T No 1428. The former Wrexham, Mold & Connah's Quay Railway terminus lies behind the building. *J. I. C. Boyd*

Below:
On the last day of service, 8 September 1962, the 6.35pm Wrexham to Ellesmere train was hauled by 0-4-2T No 1432. It had been operating these trains for over 15 years and is seen about to depart from Wrexham Central with a two-coach train. *S. D. Wainwright*

3:

Oswestry to Newtown

The first passenger train on the Oswestry & Newtown Railway departed from Oswestry on 1 May 1860, and ran a distance of 11 miles to Pool Quay. At the head of the train was the 0-4-2 mixed traffic engine No 6 *Glansevern,* built in November 1859 and eventually sold in 1867 to the Brecon & Merthyr Railway (B&M), where it became No 11 before being resold in 1881. This locomotive was one of a class of six built by Sharp Stewart, four of which were sold to the B&M.

To reduce expenditure during construction of the Oswestry & Newtown Railway, single track was installed, except for the sections between Oswestry and Llanymynech, and between Buttington and Welshpool. Bridges over the River Severn at Pool Quay and over the River Vyrnwy at Llanymynech were built to eventually take double track, incorporating an extra span. The crossing of these two rivers proved to be the main engineering problem.

During 1860, six trains ran in each direction between Oswestry and Welshpool on weekdays and two each way on Sundays, and Sunday School excursions were advertised at 1/- (5p) in covered carriages and at 2/- (10p) if travelling first-class. Whilst work was in progress over the Welshpool to Newtown section, Davies and Savin, the contractors, parted company, and Savin and Ward joined forces as Davies went on to pastures new.

The first section to open to traffic between Welshpool and Newtown was the 4-mile stretch between Newtown and Abermule. Completion of the line from Abermule to Welshpool came on 27 May 1861, when locomotive No 8 *Leighton* (which eventually passed to the B&M) hauled the first train over Kilkewydd Bridge, near Welshpool. As a point of interest, *Leighton* returned to the Cambrian with No 5 *Montgomery,* eventually being withdrawn in 1898.

Kilkewydd Bridge was rebuilt in 1921, and this was the final major engineering job carried out on the Cambrian system prior to the Grouping.

From 10 June 1861, through trains ran from Oswestry to Newtown and on to Llanidloes. Six of these left Oswestry every weekday to Welshpool, non stop, and four trains ran from Llanidloes to Oswestry (two each-way on Sundays).

By 1904, trains to Welshpool from Oswestry had increased to 10 per day, and by 1923 (in Great Western days) to only 11. In 1947, passenger workings had reduced to seven only, in each direction. Journey times were 34min by slow trains, reducing to 25min for non stop services. During 1947, auto-sets worked from Oswestry to Four Crosses and also to Llanymynech, incorporating two trailers. These were still operating in 1952.

On 19 August 1929, an inspection of 20 bridges was carried out between Oswestry and Buttington to assess to cost in bringing these up to 'red' engine strength, and during the next month an estimate of £199,000 was put forward to strengthen the permanent way between Oswestry and Aberystwyth, to enable 'red' engines to work over this line.

Great Western Mogul No 4376 was tested between Oswestry and Llynclys Junction in January 1933 in response to a request to use these heavier locomotives on ballast train workings in special circumstances. However, although permission was granted, they could not be used on the Tanat Valley line, because of weight restrictions on some bridges.

In 1938 the use of GWR 'Manors' was investigated and No 7805 *Broome Manor* was tested over the main line from Oswestry to Llynclys and from Oswestry to Gobowen. Permission to use the 'Manors' was granted.

On heading south from Oswestry towards Welshpool, the line to Porthywaen, Blodwell Junction and the Tanat Valley Light Railway branched away to the right, just before Llynclys. The station here was sandwiched between two cuttings.

After 2 miles 10 chains, Llanymynech station was reached where the Shropshire & Montgomery Light Railway joined from Shrewsbury, and the Cambrian branches to Llanfyllin and Nantmawr headed west.

From Oswestry to Llanymynech, the line gradually descended, encountering inclines of 1 in 80 and 1 in 91 en route. From here on to Buttington the terrain was comparatively flat. When the line was built, the population of Llanymynech was about 1,000, comparing with 9,000 in Oswestry, and several lime quarries and kilns were in the area. En route to Welshpool the line passed through stations at Four Crosses, Arddleen, Pool Quay and Buttington. Buttington, of course, was an important junction where trains to Shrewsbury branched away over the GW&LNW Joint Railway, via Middletown and Yockleton, the route of the 'Cambrian Coast Express'.

In September 1929, permission was requested to run 'blue' group locomotives over the stretch of track from Buttington to Welshpool with the hope that GWR 4300 class Moguls could operate – this was accordingly granted, and from 1930 these locomotives operated between Shrewsbury and Welshpool. The test run was carried out on 10 January 1930, with No 4326 checking for platform clearances. However, a 25mph speed restriction was put in force, and 10mph over the Severn Bridge, south of Buttington. In February 1930, GWR 3300 class 4-4-0 'Bulldogs' and LMS (ex-Midland) 0-6-0 goods engines were even granted permission to work over the Shrewsbury to Welshpool route, via Buttington Junction, providing the 'Bulldogs' adhered to the same speed limits as the Moguls.

However, in March 1934, due to 16½ miles of relaying of track and a mile of resleepering, 'blue' engines were forbidden over the route from Buttington Junction to Aberystwyth until the work was complete in late 1937. Further requests for differing motive power came in May 1938 when permission was granted for Class 3 (ex-L&YR) 0-6-0 'Horwich' locomotives and Class 3 2-6-2 tanks to traverse the line from Shrewsbury to Welshpool via Buttington Junction. By 1941 LMS locomotives permitted to use the line included Class 2 (ex-LNWR) 0-6-0s, Class 1 (ex-LNWR) 2-4-2Ts, ex-LNWR 19in cylinder 4-6-0 goods engines, and Class 3 (ex-L&Y) 0-6-0 Class 27 locomotives without restriction and, with the speed restrictions mentioned earlier, Class 3 (ex-L&Y), Class 28 0-6-0s, Class 3 (ex-Midland) 0-6-0s, Class 3 2-6-2Ts, and Class 6 0-8-0s.

In 1938, GWR 'Manors' were also tested over the route for the first time, and in May 1939, permission was granted to operate GWR 5100 class 2-6-2Ts over the route.

Eventually, in August 1939, after track examinations, 'blue' group engines (except 2-8-0s) were permitted, uprating the speed limit between Buttington and Welshpool from 25mph to 60mph, the only 'blue' group 4-6-0s being Great Western 'Manors' at this time. Of course, by 1958, other classes had been cleared for use, including ex-LMR 2-6-4Ts and 2-6-0 and 4-6-0 Standards.

Welshpool was the starting point for the Welshpool & Llanfair Light Railway (2ft 6in gauge) in addition to featuring a main line station with an island platform, one of only seven on the Cambrian.

The station area included coal sidings, a goods shed, a cattle dock and a water tower. The W&L station was separate, but there was an exchange shed where both standard and narrow gauge trains could transfer goods, mainly livestock. This was approached from the Oswestry direction.

The engine shed here was opened in 1861 by the Owestry & Newtown Railway, in GWR days becoming a sub-shed to Oswestry. In 1947, its allocation was one locomotive – 'Duke' No 9091. The shed building had been demolished in the early 1930s.

In July 1933, after a request to Paddington, permission was granted to operate a 19in cylinder ex-LNWR 4-6-0 goods engine to Welshpool and, following a further request to turn this locomotive on the Welshpool turntable, permission was again granted provided the engine and tender were approximately full, otherwise operation of the turntable could be difficult owing to the engine not being balanced on the table.

As the line continued south, Forden was passed, after the steady climb up from Welshpool. At Montgomery, the station was situated 1½ miles from the town, and included a 516yd loop. After a further 4 miles, Abermule was reached.

The Kerry branch joined here and this station also had an island platform. Abermule was the location where the Cambrian's worst rail disaster in 1921 occurred, related in the *Introduction.* From here the line climbed up to Newtown. The station here was one of the first to receive gas lighting, together with Machynlleth and Ellesmere. Prior to that it was lit by oil lamps. Newtown was, at first, not served by the Aberystwyth-bound 'Cambrian Coast Express', even though the up working stopped there. A slow train would take 28min to cover the route from Welshpool to Newtown, fast trains completing the journey in 24min.

The line from Oswestry to Buttington Junction closed to passengers on 23 November 1964 – 104 years after it was opened.

Below:
Showing the 'lion on wheel' emblem on its tender, 'Dean Goods' No 2538 waits at the southern end of Oswestry station with a down train. The Engineering Dept's offices can be seen in the background. The 'uncoloured A' classification can be seen on the cabside. In 1950, 13 'Dean Goods' engines were allocated to Oswestry. *E. E. Smith*

Above:
An immaculate 'Duke', No 9084 *Isle of Jersey* hauls a mixed freight through Llynclys in Great Western days. These locomotives, built in the late 1890s, had 5ft 7½in driving wheels and were the first new engines to be built with an extended smokebox. The last survivor of the class, No 9089 *St Austell* was withdrawn in July 1951, after being in traffic for 54 years. *L. N. Owen*

Below:
GWR Mogul No 6371 pauses at Llanymynech with the 12.25pm Welshpool to Whitchurch train, on 10 August 1956. The single headlamp was also used on branch passenger trains, mixed trains and certain breakdown trains, and the locomotive was allocated to Machynlleth at the time. *Brian Morrison*

Right:
Generally speaking, the 'Manors' took over duties carried out by the 'Dukes', and No 7827 *Lydham Manor*, now preserved on the Paignton & Dartmouth Railway, pulls into Four Crosses with an Oswestry train, on 26 August 1963. Note the unusual lamp on the platform. *Andrew Muckley*

Below right:
At Arddleen, a small wooden hut was the only shelter offered to passengers on its single platform. On 26 August 1963, a Welshpool train is seen approaching with Standard Class 4 No 75027 in command. *Andrew Muckley*

7827
BEWARE OF TRAINS

ARDDLEEN
75027

Buttington Junction

Above:
The mid-morning Whitchurch to Aberystwyth stopping train, hauled by Standard Class 4 No 75023, is passed by a Collett 0-6-0 heading a pick-up freight from Moat Lane Junction to Oswestry. The scene is at Buttington Junction on 4 April 1956 and the 'Cambrian Coast Express' line from Shrewsbury can be seen on the right with the token catcher in the foreground.
G. F. Bannister

Below:
With the fireman preparing to dispose of the token, 2-6-0 No 7309 arrives at Buttington Junction, on 4 April 1956, with a Shrewsbury to Aberystwyth train. Here at Buttington, in 1893, trackwork was altered, changing the two single lines to Welshpool into up and down lines. *G. F. Bannister*

Welshpool

'Dukedogs'

Above:
'Dukedog' No 9005 leaves Welshpool whilst hauling the 3.45pm Shrewsbury to Aberystwyth service, on 10 August 1956. These locomotives, prior to renumbering in the 9000 series in 1946, were known as 'Earls' — No 3205 (9005) being *Earl of Devon*. All names which had been fitted (13 in all) were removed in June/July 1937, the names being passed on to 'Castle' class 4-6-0s. No 9005 was one of the last survivors of the class. *Brian Morrison*

Below:
A fine study of 'Dukedog' No 9013 at Welshpool, awaiting departure with the 5.20pm service to Shrewsbury, on 10 August 1956. This locomotive, when numbered 3213, was allocated the name *Earl of Powis*, but never carried its nameplate. *Brian Morrison*

'Manors' at Welshpool

Left:
On a cold February day in 1965, GWR 'Manor' No 7812 *Erlestoke Manor* enters Welshpool with a morning Aberystwyth to Shrewsbury working. In the second view, the same train is highlighted by sunshine and shadows as the bitter morning picks out the escapes of steam in Welshpool's station.
J. I. C. Boyd

Right:
In failing light, on 1 January 1963, GWR 'Manor' No 7807 *Compton Manor* heads an Aberystwyth train from Welshpool. *J. I. C. Boyd*

Above:
On 14 August 1958, the 9.55am Aberystwyth to Shrewsbury train is pictured leaving Welshpool with Collett 0-6-0 No 2274 in charge. These locomotives were used on both passenger and freight duties over the Cambrian system.
Brian Morrison

Left:
A Welshpool meeting in October 1962, when Collett 0-6-0 No 2214 brought the steam crane to lift W&L engine *The Countess* back on to narrow-gauge metals, after overhaul at Oswestry. GWR 'Manor' No 7823 *Hook Norton Manor* shares the scene.
J. I. C. Boyd

Above:
The 12.27pm train for Oswestry leaves Welshpool with 4300 class Mogul No 6335 in charge, on 14 August 1958. Sidings to the left of this view accommodated a warehouse and cattle food store and, to the right, the goods shed, grain store and cattle pens were situated, in addition to the narrow-gauge W&L exchange shed. *Brian Morrison*

Below:
The daily down pick-up freight from Shrewsbury is seen shunting at Welshpool in October 1962. Note the LNWR-pattern signalbox in this view, as GWR Mogul No 7314 carries out its duty. *J. I. C. Boyd*

Pre-Grouping

Left:
Only three years after being built, Cambrian engine No 78, which became GWR No 881, is seen at Welshpool with a down passenger working in 1898. Note the interesting coaching stock.
Author's Collection

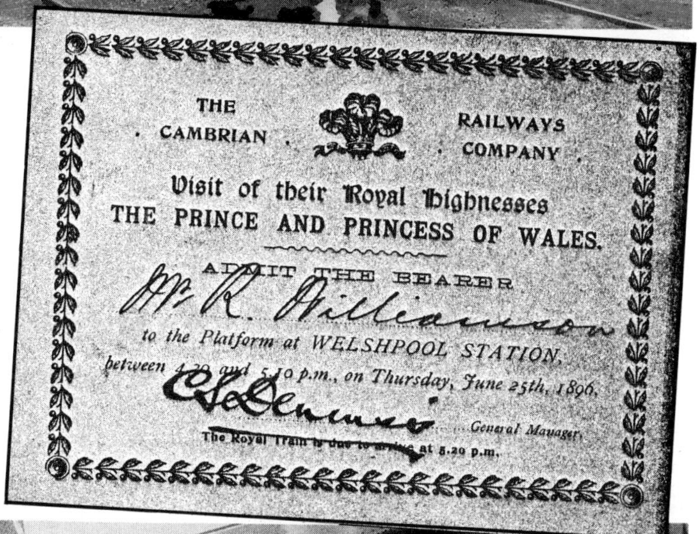

THE CAMBRIAN RAILWAYS COMPANY.

Visit of their Royal Highnesses
THE PRINCE AND PRINCESS OF WALES.

ADMIT THE BEARER
Mr R. Williamson
to the Platform at WELSHPOOL STATION,
between 4.30 and 5.10 p.m., on Thursday, June 25th, 1896.

General Manager.

The Royal Train is due to arrive at 5.20 p.m.

Right:
On 25 June 1896, King Edward VII and Queen Alexandra, when they were Prince and Princess of Wales, visited Aberystwyth, where the Prince was installed as Chancellor of the University of Wales. At Welshpool, on that day, spaces were allocated on the platform for the public to wave to the Royal couple as they passed through. This is one of the official tickets for that occasion.
Courtesy of W. J. Milner

Pre-Nationalisation

Left:
Great Western 'Duke' No 9091 *Thames* stands at Welshpool in June 1947 with an Aberystwyth to Whitchurch train. All the 'Dukes' except No 3285 (and No 3265 before reconstruction) covered over a million miles. In fact No 3254 (9054) *Cornubia* covered 1,632,815 miles. In December 1947, No 9091 comprised Welshpool's entire allocation of locomotives.
G. F. Bannister

On Shed

Above:
The engine shed at Welshpool was situated beside the down line at the Oswestry end of the station. In this view of the shed yard, ex-Cambrian engine No 99 (here as GWR No 893) awaits its next call to duty. It was withdrawn in February 1953 after 45 years of service. *E. E. Smith*

Below:
Seated on the turntable at Welshpool on 10 August 1956, is 'Dukedog' No 9013. The turntable was, in fact, situated well away from the shed building, and beside the down station platform. A water tower was also sited nearby. *Brian Morrison*

Left:
Heading the 'Cambrian Coast Express', GWR 'Manor' No 7803 *Barcote Manor* passes the signalbox situated on the platform at Forden station. Note the white-painted buffers.
Oakwood Collection

Left:
On 4 April 1956, a mid-morning Oswestry to Newtown pick-up freight passes through Montgomery with No 2538, a 'Dean Goods', in charge. *G. F. Bannister*

Below:
One of the first two 'Manors' to be allocated to Machynlleth was No 7802 *Bradley Manor*. This locomotive is seen on the main line, near Montgomery, in August 1952, with an up express, three coaches carrying roof boards.
J. F. Russell Smith

Top:
Hauling the Oswestry to Newtown morning freight, 'Dean Goods' No 2538 conveys its long train from Montgomery towards Abermule, with steam escaping from everywhere. No 2538 was the last of the class to be withdrawn, in 1957. *G. F. Bannister*

Abermule

Above:
A GWR 'Duke' hauls its eight-coach train over the main line near Abermule. Eleven of this class survived to be renumbered in the 9000 series in 1946. *N. E. Stead*

Above right:
On 4 April 1956, Collett 0-6-0 No 2239 hauls an Engineer's train from Oswestry to Machynlleth through Abermule station. The lamp positions seen here were also used on express passenger trains. *G. F. Bannister*

Approaching Newtown

Right:
Piloted by Standard 2-6-0 No 78005, GWR 'Manor' No 7802 *Bradley Manor* enters Newtown with the down 'Cambrian Coast Express' on 24 April 1957. In this 12-coach train, the last 10 coaches are the 'chocolate & cream' set. *W. G. Putnam*

2239
VIROL
VIROL

78005

4:

The Tanat Valley Light Railway

The Tanat Valley Light Railway to Llangynog opened on 5 January 1904 and was amalgamated in 1921 into the Cambrian, who had worked the 15-mile-long line, under agreement, from the start. It served an important quarrying and agricultural area but was not exactly a financial success. Differences of opinion occurred as to whether the line should be narrow gauge from Oswestry, along the Morda Valley, via Llansilin, or a standard gauge extension of the mineral branch from Llynclys to Porthywaen, via Llanyblodwel. The second option was put into practice.

The branch from Llynclys Junction to Llangynog was closed to passenger traffic on 15 January 1951, but freight still operated between Gobowen South Junction, on the old GWR Wrexham - Shrewsbury line, and Blodwell quarry, conveying Amalgamated Roadstone Corporation stone trains. The route passes through the old Oswestry station where most of the buildings still stand, and on to Llandu Junction where trains reverse into Blodwell quarry. One or two stone trains operated each week until October 1988. For the time being the line has been mothballed as the seam is not large enough to keep the line open, stone now being conveyed by road. However there is hope that a larger seam can be found at this location. If this does not materialise it is envisaged that the quarry will be closed and used as a tip for rubbish trains — reopening the line. From the line's opening in 1904, Sharp Stewart 2-4-0Ts Nos 57, 58 and 59 (GWR Nos 1192/96/97) were used, and these operated until 1927.

In May 1927, permission was first granted to use eight-wheeled coaching stock over the Tanat Valley line, but six-wheeled vehicles could not be used. Bogie stock was permitted subject to the proviso that it did not exceed the following maximum dimensions – length, 60ft; width, 9ft; height at centre, 12ft 10in; height at sides, 10ft. However, after replacement of flange rails by chaired track, in 1945, six-wheeled stock was allowed over the line, but the maximum height of vehicles was still restricted to 13ft 2in because of certain overbridges. Six-wheeled brake vans were not 'officially' granted permission until 14 March 1945. In 1939, both 50-ton and 36-ton breakdown cranes were permitted over the line provided they adhered to a speed limit of 5mph, all GWR cranes having previously being banned since the Grouping. In August 1927, engine No 1331, formerly Reading signal depot's 0-6-0T, was given the go-ahead to work over the line after being fitted with increased bunker capacity, and after the strengthening of a bridge en route. The locomotive was forbidden over private sidings on the line. No 1331 was also cleared for the Kerry and Porthywaen branches. However, in October 1928, locomotive No 574, an Armstrong 517 class 0-4-2T, was not granted permission to operate over the line, as it was heavier than those already working the route.

By April 1931, Sharp Stewart 0-6-0 No 898 (formerly Cambrian No 14) was given clearance, and by October 1943, Class 2301 'Dean Goods' locomotives could work both the Tanat Valley line, as far as Blodwell Junction, and the Nantmawr branch. This permission was granted in response to trying to find a replacement engine for No 900 on the Nantmawr branch, and Nos 2362, 2378 and 2380 were allocated for the duty, as a special wartime measure.

The three ex-Lambourn Valley 0-6-0T engines were also used on these branches, all being bought by the Cambrian in June 1904 and delivered to Oswestry. They were numbered 24, 26 and 35 by the Cambrian (GW Nos 819, 820, 821), and the last one was withdrawn in March 1946 (No 819). They were all 'ungrouped' and 'uncoloured' locomotives.

In 1904, the Tanat Valley line saw five passenger trains in each direction per week, and by 1947 this had been reduced to two such workings (one mixed) and one freight-only duty. The gradients varied from 1 in 42 to 1 in 130, and no train was allowed to exceed 25mph, with further lower-speed restrictions in many places on the line.

Left:
In August 1963, a Great Western pannier tank hauls a ballast train from Porthywaen towards Llynclys, across the A495 road from Oswestry to Llansantffraid. This crossing had to be made under supervision. Note the interesting lineside lamp. Class 31 and 37 diesels have latterly crossed the road here with ARC trains. *Andrew Muckley*

Top:
'Dean Goods' 0-6-0 No 2408 carries out shunting duties at Llangynog in September 1950. At this time, over 40 of these engines were still running — half of them on the Cambrian. *C. L. Caddy*

Above:
On 5 August 1950, 0-4-2T No 5803 prepares to leave Llangynog station with the 5.55pm train to Oswestry. The line closed to passenger traffic six months later. *G. F. Bannister*

5:

The Llanfyllin Branch

On 17 July 1863, the Llanfyllin branch opened to traffic, and, on that day, 23 packed coaches left the terminus for Borth – many of the passengers having never previously seen the sea. Many missed the return train — the journey time was 4hr. Original plans to connect Llanfyllin by a tunnel through to the Tanat Valley was discarded in favour of a route to the main line at Llanymynech. Lake Vyrnwy, situated within some of the most impressive scenery in Wales, provides Liverpool's water and when construction of the lake commenced in 1881, it was considered continuing the line to this site, a further 10 miles west of Llanfyllin, but this idea proved impracticable. Work on the lake was completed in 1910. However, the branch was used to convey the many huge water pipes and Portland cement required for the project. A team of 95 horses hauled the equipment by road from Llanfyllin to the works site – approximately 50% of these were stabled at the lake and the remainder at Llanfyllin. At one stage, in the early 1860s, there were thoughts of connecting a line from Llanfair Caereinion (now Welshpool & Llanfair Light Railway) to Llansantffraid but these were abandoned. How history might have changed regarding the preservation of the W&L!

The junction at Llanymynech was joined by the Shropshire & Montgomery Light Railway from Shrewsbury from the east. Also at this location was Rock siding, which served lime kilns. The Shropshire Union Canal proved to be a problem as trains from Llanfyllin had to reverse twice to gain enough height to pass over the canal before entering Llanymynech. Trains used a short bay at the north end of the up platform, facing Oswestry. Although this branch line rose only 240ft over its entire 9¼-mile length, gradients generally were steep, ranging from 1 in 50 around Llanymynech, to, at Llanfyllin, 1 in 221, with several banks of 1 in 75/77/80 en route.

In February 1947, permission was granted for GWR 7400 class 0-6-0PTs to operate over the branch, subject to service restrictions. This measure was taken with a view to overcoming difficulties of tender-first working during severe weather conditions. At the end of the 19th century Sharp Stewart 2-4-0T No 58 (GWR No 1196) worked the branch. In 1922, No 58, together with the other two members of the class, received considerable modifications at Swindon and returned to the Central Wales Division to work the Tanat Valley Light Railway after acquiring larger bunkers. Another early locomotive on the Llanfyllin branch was the Nasmyth Wilson 0-4-4T, a class originally designed as a 4-4-0T. Three of these were delivered in 1895 and three more followed in 1899. They were also employed on Wrexham branch trains. The last of the class of six, GWR No 10 (ex-Cambrian No 3) was withdrawn in October 1932. Their Great Western classification was 'yellow A'.

The branch closed to passengers on 18 January 1965 whilst under London Midland Region control – it did not quite make 102 years of passenger operation.

Below:
A Llanfyllin branch train returns to Oswestry on 28 August 1963 and pauses at Llanymynech, with Ivatt 2-6-0 No 46512 in charge. The waiting shelter on the opposite platform once carried 'Virol' enamel signs above the windows.
Andrew Muckley

Above:
The afternoon branch-line working from Llanfyllin to Oswestry calls at Llynclys, on 2 May 1956, hauled by 2-6-0 Ivatt No 46503. At the time, 14 of these locomotives were allocated to Oswestry. *G. F. Bannister*

Below:
Tokens are exchanged at Llansantffraid station on 9 January 1961 as Ivatt 2-6-0 No 46526 prepares to leave for Oswestry. The post that once held the oil lamps can still be seen on the platform. *E. T. Gill*

LLANFECHAIN

46510

Above left:
On 23 March 1959, the 12.40pm Oswestry to Llanfyllin train pulls into Llanfechain behind Ivatt 2-6-0 No 46510. The small goods yard can be seen in the background. *R. J. Leonard*

Left:
With its two-coach train (normal for Llanfyllin branch workings) Ivatt 2-6-0 No 46510 leaves the terminus with the 1.30pm service for Oswestry on 23 March 1959. Passenger trains always ran tender-first to Llanfyllin to return in the form shown here. *R. J. Leonard*

Above:
Ivatt 2-6-0 No 46526 arrives at Llanfyllin with the 12.55pm train from Oswestry. It is seen beside the water tower passing GWR pannier tank No 7410 which is sat under the loading gauge awaiting departure with a short freight for Oswestry, on 14 August 1959. *V. J. Vinten*

Below:
Llanfyllin — the end of the line. Running round its train for the return journey to Oswestry, Ivatt 2-6-0 No 46520 will soon take out the 1.30pm working from Llanfyllin on 4 June 1960. The brick-built goods shed is seen on the right. *H. J. Mills*

6:

The Welshpool & Llanfair Light Railway

The Welshpool & Llanfair Light Railway opened to traffic on 4 April 1903 between Llanfair Caereinion and Welshpool, after various original proposals were abandoned through objections from Powis Castle, and through impractibility, regarding a tunnel through the Berwyn mountains. Lord Powis certainly did not like the idea of a railway from Welshpool below the Llanfair Road to Sylvaen Hall being constructed when first proposed in the early 1860s, preferring a branch off the Llanfyllin line at Llansantffraid. However, the Earl's successor was its most enthusiastic promoter and the route as it stands today came into being, the line passing within half a mile of Powis Castle, about a mile from Welshpool. David Davies, after consultation with the new Earl, suggested a 2ft 3in gauge railway with provision for standard gauge if required at some later date, but when the line opened, 2ft 6in gauge was preferred.

Part of the route followed by the Welshpool & Llanfair (pronounced Klan-vire) Light Railway once carried a feeder line from Stondart quarries to the Shropshire Union Canal which was constructed around 1818. This was the Welsh Pool Rail Road; it ran for ¾ mile, and was the first railway in the world to fit rails into chairs. The line ceased operations in 1854.

The two 0-6-0T locomotives used on the W&L line, and still operating after 88 years, were built by Beyer Peacock, and carried Walschaerts valvegear. They were maintained at Oswestry.

Passenger services ceased in February 1931 but freight traffic continued until 1956. The idea of preserving the W&L first came in 1952, and William Morris founded a preservation society in November 1956. However, the local authority strongly objected to the line still continuing into Welshpool, and although British Railways stated that it was prepared to negotiate transfer of the line to the preservation group, this was only on the basis that the line was terminated west of Raven Square, beyond the point where the A458 trunk road had been crossed by the narrow-gauge line into Welshpool. This situation remains today.

On 28 July 1961, *The Earl* was steamed on the line, after a lay-off of 4½ years. *The Countess* returned home in October 1962, and the line reopened to passengers on 6 April 1963. On 17 August 1963, the last train ran into Welshpool, double-headed, as the local council had purchased the land east of the A458 crossing. A stopblock was constructed west of Raven Square early in 1964 with a loop.

Severe flooding, seriously damaging the track near Dolrhyd Mill and the viaduct over the River Banwy, caused termination of trains from Llanfair at Heniarth, 1¼ miles from Llanfair Caereinion in September 1964. Following eight months of work by the Royal Engineers on the bridge and track, trains continued to Castle Caereinion in August 1965. Owing to deterioration of track, maintenance was a long job, and expensive, and funds were hard to come by, but work continued.

In 1969, a weedkilling train ran to Raven Square, and by 1972 passenger services again ran to Sylvaen, after a lay-off of eight years. Now, happily, steam-hauled trains run all the way to Raven Square – a remarkable achievement. Because the scope of this book is confined to the period when steam on the Cambrian system was in everyday service, I have chosen to include photographs of the W&L in Cambrian/GWR/BR days only, prior to preservation, a period not covered so prolifically in present-day publications (with no disrespect to the W&L Preservation Society which is doing a tremendous job).

To go back in history, in 1904 there were five trains in each direction per day, but none on Sundays, the journey time over the 9-mile route taking 1hr 10min. The line was worked by the Cambrian from the start, although the W&L was an independent company, obtaining 60% of the gross receipts. By 1923, trains had reduced to four in each direction, the journey being completed in 1hr 5min. In the days of 'freight only', the 1947 service timetable shows, in each direction, one Mondays Only working, one freight operating on Tuesdays, Thursdays and Fridays, and also a livestock train on Mondays only. The line was worked by wooden train staff – one engine in steam. Line speed was restricted to 15mph, and less in certain areas.

In 1925, livestock movement over the line totalled 344 wagonloads, and milk and timber traffic had increased from previous years, but coal haulage was down. Load transfer from standard gauge to narrow-gauge wagons took place at Welshpool in the 'Exchange Shed'. The engine shed at Welshpool, built to house the narrow-gauge locomotives, was of corrugated sheeting over a steel framework, with a curved central raised ridge, and was attached to a longer building – the carriage shed. The engine shed was glazed along the outer wall of the 55ft-long building – it was 10ft wide.

Along the route from Welshpool to Llanfair Caereinion gradients varied from 1 in 300 to the very steep 1 in 29, ¾-mile stretch, between Raven Square and Golfa. We look forward to steam continuing over this picturesque line for many years to come.

Facing page, bottom:
In this view photographed before World War 1, W&L 0-6-0T No 1 *The Earl* is pictured at Llanfair Caereinion with two early first/third-class composites on a mixed train. Note the gated platform ends on the coaching stock.
Ian Allan Library

Left:
Standard-gauge cattle trucks stand adjacent to W&L narrow-gauge stock at Welshpool in Great Western days, as No 822 *The Earl* leaves for Llanfair. The W&L engine shed with its roof vents can be seen in the background beside the carriage shed.
Ian Allan Library

Left:
No 822 *The Earl* prepares to leave Llanfair with loaded cattle trucks in GWR days. The letters 'GWR' appeared on the numberplate above the number.
Real Photographs

Right:
No 823 *Countess* is pictured at Welshpool on 6 April 1926. Originally named *The Countess*, the definitive article was dropped by the GWR as the full name would not fit in its new position on the cabside.
LCGB Ken Nunn Collection

Below:
On 10 August 1956, No 822, *The Earl*, now having lost its nameplate and carrying its 89A shedplate, arrives at Welshpool with a freight.
Brian Morrison

Above:
A scene from Welshpool, where the narrow-gauge track ran over a brook and very close to living accommodation. Extreme care needed to be taken here. *Montgomery Express*

Below:
On 21 April 1955, No 823 enters Welshpool with the daily freight from Llanfair and crosses the main A458 road at Raven Square. *G. F. Bannister*

Right:
No 822 heads the daily freight from Welshpool, on 4 April 1956, and approaches Golfa, on its way to Llanfair.
G. F. Bannister

Below:
On 10 July 1956, No 822 climbs out of Welshpool with the daily freight consisting of three wagons of coal, one of ash and one wagon of fertiliser. Powis Castle is advertised in the field alongside the line. *T. K. Widd*

7:

The Kerry Branch

The branch from Abermule to Kerry, some 3¾ miles in length, passed through the Mule Gorge and approached Kerry up a 1 in 43 incline. On 2 March 1863, the first engine, hauling one carriage, arrived at Kerry. It was intended to attach a wagon of coal at Abermule for distribution amongst the poorer residents at Kerry but this proved too much for the locomotive – this was conveyed on the following day.

Initially, the Kerry branch operated two passenger and two mixed trains daily, each way, with an extra passenger working during the summer, and on Mondays, Tuesdays and Saturdays. The journey time was 25min. Second-class passenger accommodation ceased on the Kerry branch by the 1880s, trains being restricted to first and third-class carriages. Only a few years later, second-class accommodation was abolished altogether.

By 1904, services had more or less doubled, but by 1923 passenger services consisted of one train on Wednesdays, Fridays and Saturdays, two on Mondays and Thursdays, and three on Tuesdays – still no Sunday service and Goitre and Fronfraith were just request stops.

The Kerry branch's passenger services ceased on 9 February 1931, but freight continued, including livestock traffic for the Kerry sheep sales. In 1904, dead game, poultry and rabbits were conveyed by passenger train from Kerry to Birmingham at a rate of 2/6d per ½cwt, and meat was taken to London at 4/- per ½cwt.

By 1947, one freight train per day (Mondays excepted) travelled from Abermule to Kerry, departing at 9am, remaining at Kerry for 1¼hr before returning. The line was worked by wooden train staff and double-heading over the route was allowed. The branch was closed to freight on 1 May 1956.

Early locomotives working on the Kerry branch included the Sharp Stewart 0-4-0STs built in 1863, numbered 36, 37 and 38, and named *Plasfynnon*, *Mountaineer* and *Prometheus*. They were all withdrawn by 1907. After purchase of the three Lambourn Valley Railway locomotives by the Cambrian in 1904, these engines were used on the Kerry branch in addition to the Tanat Valley, Van, Nantmawr, and Porthywaen lines. In 1927, even the former Reading signal depot's 0-6-0T, No 1331, was passed for working the branch. In August 1928 the Sharp Stewart 4-4-0s has been passed for operating Kerry trains during livestock sales and No 1118 of 1886 had come on to the line. Concern was expressed over the damage that this locomotive was causing the permanent way and this led to it being withdrawn from the line. Despite requests to remove the engine from the branch, No 1118 was still found to be operating on 31 October 1928 and instructions were again given to remove it immediately or the branch would close to traffic. Requests were made again to use this type of locomotive in August 1929 and permission was granted, but for use only on two days – 6 and 13 September.

In 1928, Class 517 locomotives were refused permission to work the line, owing to their weight and, in September 1929, Class 2021 0-6-0PT No 2075, working the Kerry branch at the time, was reported cutting the running face of the rails at Abermule, and on the curves of the branch line, and an examination of the locomotive's wheels was ordered. In August 1939 0-6-0T No 819 (ex-Cambrian No 24) and 0-6-0PT No 2068 worked the line, and operated from Moat Lane shed.

Regarding rolling stock, eight-wheeled carriages were allowed from May 1927, but six-wheeled stock was forbidden. However, dimension limitations were 56ft × 8ft 6in, 50ft × 8ft 9in, and 45ft × 9ft. Both 50-ton and 36-ton breakdown cranes were subject to 5mph over the Kerry branch.

Left:
The last working 'Dean Goods' 0-6-0, No 2538, is seen in the Kerry branch platform at Abermule about to leave for Newtown on 4 April 1956, with a freight.
G. F. Bannister

31st October, 1928.

48271/7. B.

O. 9101.

Dear Sir,

ENGINES WORKING ON KERRY BRANCH.

In August last I gave authority for an engine of the 1112 type to work specially on the Kerry Branch during the Sheep sales, but as engine 1118 has been working daily on this Branch for some time and is causing considerable damage to the permanent way I wrote to Mr Warwick on the 19th instant asking him to arrange for this to be withdrawn at once. I find this engine is still working, and unless it can be withdrawn immediately I shall have to consider closing the Branch to traffic.

Yours faithfully,

E. COLCLOUGH, ESQ.

Left:
A letter from the Chief Mechanical Engineer's Department to the District Traffic Manager's Office at Oswestry, indicating the concern at the continued use of No 1118 on the Kerry branch.

Below:
In a silvan setting, 'Dean Goods' No 2538 passes through the Mule Gorge, on 4 April 1956, on its return from Kerry with the thrice-weekly freight.
G. F. Bannister

Right:
Cambrian 4-4-0 No 21, which in its later guise as GWR No 1118 caused so much consternation on the Kerry branch. *LPC*

Below:
On 2 December 1955, 0-6-0 No 2516, now preserved at Swindon, climbs through the Mule Gorge with the Kerry to Abermule freight, and is seen north of Middle Hill. Note the number also appears on the buffer beam. *G. F. Bannister*

Above:
'Dean Goods' No 2538 carries out leisurely shunting duties in Kerry station yard on 4 April 1956. *G. F. Bannister*

Below:
On 2 December 1955, 'Dean Goods' No 2516 awaits departure from Kerry station for Abermule with the thrice-weekly freight. The brake van (No 68710) was branded 'ABERMULE' — a name removed in June 1956 after closure of the branch.
G. F. Bannister

8:

Newtown to Llanidloes

The Llanidloes & Newtown Railway was the first of the Cambrian-absorbed companies to be completed, in April 1859. Locomotives associated with the line, and believed to have been owned by David Davies, responsible with Thomas Savin for the line's construction, were *Dove, Squirrel, Llanidloes, Llewelyn* and *Llandinam,* and *Llewelyn* hauled the line's first passenger train on 31 August 1859. *Dove* was delivered by road from Oswestry to Llandinam and was a secondhand 2-2-2 locomotive. *Llandinam* was an 0-6-0ST, delivered by the same route, and eventually went to the Pembroke & Tenby Railway. *Milford,* one of four 0-4-2STs, built by Sharp Stewart for Davies and Savin, accompanied *Llandinam* on the inaugural run. On completion of the L&NR, *Milford* was sold, taken over by the Cambrian in 1864, and used on the construction of the Aberystwyth & Welsh Coast Railway, after which it went to Oswestry for yard shunting duties and Porthywaen quarry workings. The 'Black Donkey', as it was known, No 1 *Enterprise,* was delivered in 1858 to work the L&NR, and later was employed during the construction of the Llanfyllin branch.

Originally, the line from Llanidloes to Newtown was planned as double track, but this was changed to single with provision for doubling by building bridges accordingly.

Freight trains commenced operations on 30 April 1859 and regular passenger services came into force in September of that year. The first through train from Oswestry to Llanidloes ran on 10 June 1861, although four trains ran in each direction between Newtown and Llanidloes in 1860.

In January 1863, the line became a branch when it was joined to the Machynlleth to Oswestry line at Moat Lane Junction. A through station at Llanidloes was completed in readiness for the Mid-Wales Railway early in 1864, and the old L&NR terminus there was closed.

The first station, on leaving Newtown, was Scafell, built after the through line to Machynlleth was completed. Scafell was 312ft above sea level, comparing with Llanidloes at 533ft, and its station was only a conditional stop. It was later goods only.

The climb from here to Moat Lane was gradual and, in the original timetables, the station in this vicinity was named Caersws. An L&NR shed was built here in 1859, and in 1957 a corrugated-iron building replaced the original timber shed of the same dimensions – 35ft × 70ft. Its turntable was moved from the Oswestry end of the station to a position close to the shed building. In August 1898 Moat Lane's allocation was three locomotives – 2-4-0T No 1 (GWR No 1329), 2-4-0 No 10 (GWR No 1328), and 4-4-0 No 65 (GWR No 1096). In December 1947, its allocation consisted of ex-Cambrian 0-6-0 No 844, three 'Dean Goods' engines, Nos 2354, 2449 and 2482, and 0-6-0PT No 2068. The depot closed in 1962.

From here, the line climbed to Llandinam and on to Dolwen. Further inclines, the most severe being 1 in 132, were to be found over the route to Llanidloes. The line joined two important flannel-making centres, Newtown and Llanidloes and, in 1874, there were three passenger trains in each direction, one mixed train, one mail train, and three freight workings. Moat Lane Junction had appeared in the timetables since 1863. At Llanidloes station, a refreshment room was in operation by 1872, one of a series introduced by the Cambrian, despite the infrequency of passenger trains, whereas at Moat Lane, in 1871, passengers had to put in a petition to have a drinking fountain installed.

The second of the L&NR locomotive sheds was constructed in 1859 at Llanidloes, also closing in 1962. Situated at the north end of the station, the two-road brick and slate building measured 35ft × 75ft, and a turntable was provided. The Llanidloes allocation in 1898 was 4-4-0s Nos 20/60/63/71 (GWR Nos 1117/1112/1091/1104), 4-4-0Ts Nos 2 and 33 (GWR Nos 1129 and 1131), 2-4-0T No 57 (GWR No 1192), and No 18 (not passed to GWR). In December 1947, the allocation comprised a 2251 class 0-6-0 No 2255 and three 'Dean Goods' engines, Nos 2382, 2516 and 2569.

On 30 July 1912, permission was granted to double the line from Newtown to Moat Lane. The distance involved was 4 miles 49 chains, and the steepest gradient was 1 in 150. The fences were the original hedges and no extra land was required for the work to be carried out. Initially, a 25mph speed limit was enforced over the new line.

In 1932, the Llanidloes auto-train worked to Moat Lane (Set No 58), after earlier in the day working south from Llanidloes to Builth Wells, departing there to arrive at Moat Lane around midday, eventually taking the same route back to Llanidloes.

Between Moat Lane and Llanidloes 'yellow' group locomotives were permitted from December 1940, and only certain 'yellow' engines (*detailed in the next section*) were permitted beyond Llanidloes, as far as Brecon, but generally this was 'uncoloured' territory. Passenger traffic ceased from Moat Lane to Llanidloes and Brecon on 31 December 1962, but the section from Moat Lane to Llanidloes remained open as a siding for cement trains from Aberthaw in South Wales, via Shrewsbury, until 2 October 1967 – that section of track was finally lifted in March/April 1969. The cement was required during the building of the Clywedog Dam, 3 miles from Llanidloes.

75026
VIROL

78005

Left:
Standard 4-6-0 No 75026 enters Newtown with the 12.45pm Aberystwyth to Whitchurch and Shrewsbury service on 24 April 1957. At the time, only Nos 75005/06/20/26 of this class were allocated to Oswestry.
W. G. Putnam

Below left:
Class 2 Standard 2-6-0 No 78005 waits in the bay at Newtown and prepares to leave with the 4.20pm local train to Machynlleth. At this time, around 1957, the first 10 engines in this class were allocated to the Western Region — six to the Central Wales Division.
Andrew Muckley

Above:
Trains cross at Newtown in August 1963, as ex-GWR Mogul No 6368 awaits departure for Aberystwyth with an evening service, and shares the station with an eastbound working. This was one of only four ex-GWR Moguls allocated to the Central Wales Division in 1963, although ex-LMS Moguls numbered 22, and Standard Moguls, four.
Andrew Muckley

Right:
'Dean Goods' No 2343 leaves Moat Lane Junction with a four-coach Mid-Wales line train. This engine was one of the six 'Dean Goods' 0-6-0s allocated to Brecon in 1950.
Ian Allan Library

78002

7803

Above left:
In October 1962, a Brecon train, hauled by BR Standard Mogul No 78002, enters Moat Lane Junction from the north. These Darlington-built 2-6-0s were designed by Riddles and derived directly from the Ivatt Moguls. *J. I. C. Boyd*

Left:
Double-headed power at Moat Lane Junction (c1956). 'Dukedog' No 9009 pilots GWR 'Manor' No 7803 *Barcote Manor*, one of the first to be allocated to Oswestry.
R. S. Carpenter

Above:
Another Brecon-allocated 'Dean Goods' 0-6-0, No 2351, is seen between Dolwen and Llanidloes on a down working with mixed coaching stock at milepost 58¾ (miles from Whitchurch). *E. E. Smith*

Below:
Llanidloes, with its fine station building, on 4 August 1962, less than five months before the line closed to passenger traffic. Ivatt 2-6-0 No 46511 is seen preparing to leave for Moat Lane Junction with the 8.05am train. *E. Wilmshurst*

9:

Llanidloes to Talyllyn Junction

The Mid-Wales Railway, as the Llanidloes to Talyllyn Junction line was known, opened to freight on 1 September 1864 and to passenger traffic from Three Cocks to Talyllyn Junction on 19 September and from Llanidloes to Three Cocks on 21 September 1864. The station at Talyllyn Junction was, in fact, a Brecon & Merthyr Railway station.

Amalgamation of the Mid-Wales Railway into the Cambrian Railways system came in 1904 — the Cambrian had operated the line since 1888.

Between Llanidloes and Tylwch was Pentpontbren Junction (no station). This was the point where the Manchester & Milford Railway joined from Llangurig. This section of the M&M was lifted in 1883 with no scheduled train ever running over it. On 1 November 1866, the LNWR Central Wales line reached the Mid-Wales Railway at Builth Road where a curve joined the lower-level Mid-Wales line to the high-level LNWR. Ownership of the adjoining spur transferred to the LNWR on 4 July 1870.

The route from Llanidloes to Talyllyn Junction included five stations opened by the GWR — Glan-yr-Afon, Marteg, Llanfaredd, Llanstephan, and Llangorse Lake. Builth Road was named Llechryd until 1889 and from 1912 Boughwood was named Boughwood & Llyswen. Talyllyn's Mid-Wales Railway station closed in 1878. At Three Cocks Junction, the Midland Railway joined from Hereford and Hay-on-Wye and, in fact, the first 29 chains of the Hereford line was Cambrian-owned. The Midland had running powers from here to Talyllyn Junction and Brecon. Llanidloes station (c1943) had up and down refuge sidings — the down neck would take one locomotive, 32 wagons and a brake van, and the up neck, one locomotive, 50 wagons and a brake van. The up loop was 684ft long and the down, 693ft. In the 1920s, special trains were organised for the conveyance of livestock to the Llanidloes Fair, with livestock being loaded into MEX cattle wagons and attached to passenger trains.

The climb towards Tylwch was very steep for the first part of the journey, 1 in 75 for almost a mile, and after Glan-yr-Afon, Pantydwr was reached, the highest station on the Cambrian, 947ft above sea level, and reached by an incline of 1 in 77. From this point the line descended towards St Harmons and Marteg Tunnel, 372yd long, near where another GWR-built station was constructed. Rhayader was the next station on the route, another location where livestock fairs took place. Rhayader station was constructed in stone with a signal cabin situated on the end of the platform where a water column was originally sited. The station had an island platform and all platforms were low-level and, within the yard, there were three places where locomotives could run round their trains — a 4-ton crane was located in one of the sidings and one of the through sidings passed through a granary. A corn mill was situated adjacent to the station. Cattle pens and stables showed the importance of this rural location.

In July 1904, the Royal Train visited Rhayader, conveying the late King Edward VII and his Consort to open the Birmingham Waterworks in the Elan Valley nearby. This is reputed to have been the only train to have used the spur joining the Cambrian line with the LNWR at Builth Road, during pre-Grouping days. Proposals for lines from Rhayader to Aberystwyth, Cross Gates and Yspytty Ystwyth never came to fruition.

After passing through Doldowlod and Newbridge-on-Wye, Builth Road was reached, where the LNWR line passed over from Carmarthen to Craven Arms. Builth Road station was sited on the Three Cocks side of the LNWR line with its High Level station, which passed over the Cambrian route here, and a luggage lift conveyed items between the two stations.

Originally, Builth Road station comprised only one platform with a timber waiting shelter, but by 1893 an additional platform was added, together with a passing loop, for up trains — station buildings were then constructed on the down platform. Passengers crossed the line by using the barrow crossing. The words 'Low Level' were not added to the nameboards until 1950. Two signalboxes existed here — North and South.

At Builth Wells, a small engine shed operated from 1864, closing in 1957. Its allocation in 1898 comprised one 4-4-0, No 64 (GWR No 1093) and two 0-6-0s, Nos 29 and 43 (GWR Nos 849 and 873). In December 1947 two locomotives were allocated, both 0-6-0s, Nos 895 and 2556. It was a single-road shed and a turntable was situated nearer the station and adjacent to the goods shed. The engine shed was south of the station.

Sixteen miles from Llanidloes, Three Cocks Junction (305ft above sea level) was reached, where Midland engines joined the Cambrian, en route to Brecon. The route to Three Cocks passed through Erwood (pronounced Errood) and Boughwood (pronounced Bockrood).

The final stage of the journey to Talyllyn Junction was a steady climb, where gradients of 1 in 75 were encountered. At Talyllyn, 610ft above sea level, where the junction to Brecon and Dowlais was formed, the first 16 chains after the junction on both lines were, in fact, Cambrian-owned. From there on, the Brecon & Merthyr Railway took over. The railway here formed a triangle, the lower line forming a through B&M route from Dowlais to Brecon.

Originally, in 1864/5, six Kitson 0-4-2s and six Kitson 0-6-0s were used on the Mid-Wales Railway. The first 0-4-2, No 1, was named *James Watt*, a name carried until 1874. In 1873, two Sharp Stewart 0-6-0s, originally ordered by the Furness Railway, were purchased. By 1893, owing to increased traffic, the first of 22 new 61 class locomotives came into use — 16 built by Sharp Stewart, two at Oswestry Works and four by R. Stephenson & Co over a period from 1893-1904. These engines operated on the Mid-Wales line from the summer of 1905 and were the heaviest allowed over the route.

By 1925, the 2-4-0 3201 and 3501 'Stella' class engines were permitted, and also the 2-4-0 3232 class, 4-4-0 3521 class and 0-6-0 2301 class 'Dean Goods' engines, as they were all in the 'uncoloured' group. This excluded 'yellow' group 'Belpaires' until 1940 when the line was passed for 'yellow' working. Three Cocks Junction to Brecon was a 'blue' route. From September 1926, Moat Lane to Three Cocks was limited to a 14-ton axle load. A request to use 'yellow' engines Nos 3521, 3545 and 3554 was turned down in February 1930 and a misunderstanding over the use of GWR Class 5100 2-6-2Ts between Three Cocks and Brecon arose when a 10-year-old authority was deleted from the engine book in August 1939

In December 1940, 'yellow' engines Nos 844/849/855/864/873/887/892/894/895/896 were permitted to work between Moat Lane and Brecon. Another Royal Train travelled from Shrewsbury to Rhayader in October 1952 and a request to use double-headed 'Manors' from Moat Lane to Builth Wells and return was granted subject to clearance tests. During 1913/14, four trains over the Mid-Wales line were delayed 'due to poor coal', it was claimed. The locomotives involved were Nos 73, 75, 76 and 80.

The last booked train to run over the entire Mid-Wales section was on Saturday 29 December 1962 and the last through Moat Lane return working was the 9.55am Moat Lane to Brecon, and the 1.20pm Brecon to Moat Lane return journey, hauled by ex-LMS 2-6-0 No 46516. This class of locomotive had worked the Central Wales Division since March 1953.

Below:
Ivatt 2-6-0 No 46508, with the 9.55am Moat Lane to Brecon train, stops at Tylwch Halt on 29 July 1959. Note the abandoned second platform on the right and the site of the original crossing loop. *R. O. Tuck*

849
849

2538

Left:
A classic Cambrian scene as 0-6-0 No 849, built in 1918 (ex-Cambrian No 29) is seen near Tylwch with a Moat Lane to Brecon train in August 1952. No 849 was withdrawn in October 1954, having been fitted with a non-superheated boiler in April 1931 which was eventually replaced with a superheated version in August 1937. It reverted to a non-superheated boiler in April 1947 and back again to a superheated one in October 1951. *J. F. Russell Smith*

Below left:
The wooded countryside near Rhayader sets the scene for 'Dean Goods' No 2538 with a northbound freight on 9 October 1951. *J. F. Russell Smith*

Above:
A two-coach train for Moat Lane leaves Builth Wells on 1 August 1962 with Ivatt 2-6-0 No 46518 in charge. At this time, all engines in this class, from Nos 46503 to 46524 were allocated to Oswestry, except Nos 46506 and 46517 which were at St Philip's Marsh (82B). *P. J. Lynch*

Below:
In this beautiful rural setting beside the River Wye, the 9.55am Moat Lane to Brecon train, headed by 2-6-0 No 46516, leaves Llanfaredd (hidden by smoke), on 28 May 1962, southbound. *J. Spencer Gilks*

Above:
A southbound train, headed by ex-Cambrian 0-6-0 No 855 is pictured near Erwood on 10 October 1950. This locomotive (ex-Cambrian No 31) was built by Beyer Peacock in 1919.
J. F. Russell Smith

Below:
On 24 May 1956, No 46516 is seen at Erwood — a fine study of a popular Swindon-built locomotive. On this stretch of line between Three Cocks Junction and Llanidloes, a June 1921 notice forbade double-heading under any circumstances.
D. Fish

Left:
Brecon-allocated 'Dean Goods' No 2401, hauling corridor stock, a luxury brought to the Brecon line by the GWR, is seen on a northbound train near Erwood on 11 October 1950.
J. F. Russell Smith

Below:
Three Cocks Junction, showing three trains having recently arrived. The time is 1.55pm on 29 September 1962, and (left to right) we see the 1.20pm Brecon to Moat Lane train hauled by No 46519; the 12.30pm Builth Road Low Level to Brecon with No 46514 in charge; and No 46523 hauling the 12.42pm Hereford to Three Cocks Junction train — a connecting service. The Ivatt Moguls had taken over!
P. W. Simmonds

Above:
With a Brecon to Hereford train, ex-L&YR 0-6-0 3F No 12143, introduced in 1889, with open cab, is seen (c1935) at Three Cocks Junction. These Aspinall locomotives lasted well into the 1960s. *Ian Allan Library*

Below:
A short freight, hauled by Oswestry-based 'Dean Goods' No 2482, is seen near Talgarth on 11 October 1950. This was one of the 13 Oswestry-based members of the class at the time. *J. F. Russell Smith*

Bottom:
Another 'Dean Goods', No 2538, heads a two-coach Moat Lane to Brecon train, near Talgarth, on 11 October 1950, sporting the 'lion on wheel' emblem. *J. F. Russell Smith*

10:

Moat Lane Junction to Machynlleth

This route, which opened to passenger traffic in January 1863, peaked at Talerddig Summit, 693ft above sea level, with gradients as steep as 1 in 52 approaching from the Llanbrynmair direction.

The first part of the line to be worked upon was the section from Moat Lane to Talerddig, early in 1859, and was 9 miles in length. Later that year work began on the heavy-inclined Talerddig to Llanbrynmair stretch, but owing to the rugged landscape and wet weather, this was slow work. The heavy rainfalls caused landslips in Talerddig Cutting and at Pontdolgoch, where there was also a cutting. Originally, it was to be 'Talerddig Tunnel' (not cutting), but this proved to be too expensive and, at the time the cutting here was the deepest rock cutting in the world with a depth of 120ft. The workforce even found a little gold in the rock. The River Carno also had to be diverted and made deeper to prevent flooding on the line. Steep gradients required heavier track, and double-headed rail on cast-iron chairs, where elsewhere flat-bottomed rail was used.

From Moat Lane Junction, the line travelled west to Caersws, where the branch line to the Van lead mines left the main line – a branch which finally closed in 1940. The Van branch was first worked by two Manning Wardle 0-6-0 saddle tanks – one named *Caersws*. In later years (c1939), Manning Wardle 0-6-0STIC No 824 (ex-Cambrian No 30) was used, together with (on occasions) No 819, a Hunslet 0-6-0T and GWR 0-6-0PT No 2068. More information on the Van branch can be found in the *Introduction*. Caersws station was built on the site of a Roman settlement.

In 1868, during that winter's fierce storms, the River Severn and its tributaries rose so quickly that the driver of a freight train from Machynlleth to Newtown found water lapping over the footplate as he descended from Talerddig, past Carno and on towards Caersws. The water was almost putting out the fire in the firebox. However, on the return journey, on warily passing across the river bridge at Caersws, the locomotive suddenly toppled over the embankment adjoining the bridge where earthworks had been washed away, although the bridge itself had remained secure. Sadly, both driver and fireman were killed in the incident.

A Cambrian director was sent to the scene from Oswestry, and travelled on a packed passenger train which was stopped just in time after it was noticed that another bridge, at Pontdolgoch was found to be giving way under the fierce torrent.

Carno, between Pontdolgoch and Talerddig, was the scene of an amusing incident, involving some presence of mind, during the early days when railway excursions to the coast at Borth were constantly packed to capacity. Carno station ran out of tickets and the stationmaster, an Irishman, not wishing to lose revenue, booked around 40 farm labourers with 'cattle tickets'.

Talerddig, very vulnerable to the elements, suffered very severe weather conditions and landslides such as in the tragic January 1921 incident. Fortunately certain landslides did not result in fatalities, passengers only being inconvenienced by having to be transported, together with their luggage, by road for a few miles, whilst the line was cleared.

The climb to Talerddig Summit proved to be one of the most popular photographic spots for railway enthusiasts, as powerful pictures were made possible as double-headed trains pounded their way to the top. The ex-Metropolitan 4-4-0Ts were tried on banking duties to the summit but proved unsuccessful and over the years various combinations of double-headed power were to be seen, such as 4500 class tanks with 'Jones Goods' and trains were sometimes banked at the rear by GWR 'Dukes'. In later days, of course, the 'Manors' and Standards were regulars on the bank, the Standards often in tandem.

From Talerddig Summit it was all downhill to Machynlleth. After passing Llanbrynmair, the line passed through Commins Coch Halt and Cemmes Road, where the branch from Dinas Mawddwy joined, originally installed chiefly for mineral traffic. From here the main line continued a further eight miles to Machynlleth. Journey time from Newtown to Machynlleth was about 55min for slow trains and 44min for fast.

In 1874, trains in each direction totalled three passenger, one mixed, one mail, and two goods, and by 1923, passenger workings had reached 12, with trains from Manchester, Liverpool, Birmingham and London.

Machynlleth operated a comparatively large allocation of locomotives from its depot, which opened in 1863. The shed building housed three roads, one which terminated within, and a feature of this location was the sheer rock face which ran along the back of the shed yard and cattle pens. The manually-operated turntable was a favourite with all enthusiasts. Machynlleth locomotive allocations (1948 and 1959) are shown in the *Introduction* but, in comparison, at the Grouping, 1922/23 allocations included 'Barnum' No 3210, with Nos 3207/13/17/25 seeing out their lives there in the mid-1930s. Two 'Stella' class engines, Nos 3501/19 lasted at Machynlleth until 1930 and 1925 respectively.

In 1939 the GWR 4300 class Moguls entered the Central Wales Division, three being based at Machynlleth. Ex-GWR 2-4-0s Nos 1328 and 1329 (ex-Cambrian Nos 10 and 1) worked from this location until withdrawn in 1926/27, and in 1930, three 3521 class 4-4-0s were allocated to Machynlleth and shunting was carried out by 850 class pannier tanks.

The station at Machynlleth opened on 3 January 1863 and was built in large stone blocks. The station area also comprised the Corris Railway station (2ft 3in gauge) which opened in 1859 to convey slate from Corris, via Machynlleth to Derwenlas on the Dovey estuary. The line was originally worked by horses, and it was never Cambrian-owned. The GWR purchased the Corris Railway in August 1930, and it closed to passengers, conveyed since 1883, in 1931. Freight traffic ceased in August 1948.

Plans by the GWR to carry out certain alterations at Machynlleth soon after takeover did not materialise owing to cost, and lack of space. In 1867, gas lighting reached Machynlleth station, replacing lamps lit by oil – firstly by seal oil and then mineral oil.

Prior to the Grouping, the longest of all Royal rail tours undertaken by a reigning monarch, was in July 1911, when King George V, Queen Mary and other members of the Royal family travelled from Caernarvon to Machynlleth, via Afon Wen and the Cambrian Coast route for the laying of the foundation stone of the Welsh National Library at Aberystwyth. Two days later, the Royal Train travelled from Machynlleth to Whitchurch, on its way to Scotland.

Above:
The Mid-Wales line auto-train, comprising 'Dean Goods' No 2543 and auto-trailer No 118, stands in the platform at Moat Lane Junction — a fascinating picture of this unusual train. *E. E. Smith*

Caersws

Right:
The 4.25pm Newtown to Machynlleth local arrives at Caersws, on 19 August 1963, hauled by 'Mickey Mouse' No 46446. This engine had recently been transferred to Machynlleth depot from Rugby. Token exchange is about to take place. *M. J. Fox*

Above:
With another token exchange imminent, Standard 2-6-4T No 80131 steams into Caersws station with the 2.45pm Aberystwyth to Oswestry train, on 27 August 1964.
Andrew Muckley

Below:
Pontdolgoch station, and an afternoon school special arrives in August 1964 with Standard 2-6-2T No 82005 in charge. By 1965, all Machynlleth-allocated Standard 2-6-2Ts had been transferred to Nine Elms and Patricroft. *Andrew Muckley*

Above:
This view of Carno station shows the fine stone-block buildings, as the token is about to be exchanged from 'Manor' No 7828 *Odney Manor*, seen with a stopping train for Shrewsbury, on 2 September 1963. *Andrew Muckley*

Below:
Hauling a variety of coaching stock, 'Earl' No 3202 descends Talerddig Bank with a down train, prior to nationalisation. *E. E. Smith*

Talerddig

Left and below:
An impressive Saturday spectacle as the Aberystwyth to Shrewsbury section of the London train's run at Talerddig makes heavy going for the locomotives in September 1966 (left). The train is seen going away (below) through the inclined cutting under the power of two BR Standard 4-6-0s, Nos 75052 and 75053 (leading), creating a powerful sight. *J. I. C. Boyd*

Above:
Double-headed 'Dukes', a fine sight, No 3276 *St Agnes* (originally No 3287) leading on this occasion. 'Duke' pairings were a common occurrence on Talerddig Bank in the late 1920s and early 1930s. This pair are seen descending the bank with a nine-coach train at a favourite photographic location. *St Agnes* lost its nameplate in August 1930 on instructions from the Traffic Department who claimed that passengers were mistaking the locomotive names for train destinations. *Author's Collection*

Below:
At the same location, again in GWR days, double-headed 'Earls' with No 3214 leading. Apart from being found on the Cambrian, these 4-4-0s were allocated to Bristol, Swindon, Oxford and Didcot *N. E. Stead*

Right:
On 6 August 1966, Standard 4-6-0 No 75006 prepares to ascend Talerddig Bank with help at the rear from Standard 2-6-0 No 76040, as it hauls its Birmingham to Aberystwyth train. *Derek Cross*

Below right:
With the token for the section to Carno, Collett 0-6-0 No 2255 passes Talerddig station with the 8.12am Barmouth to Birmingham train on 22 August 1959. No 2255 was paired with an ex-ROD tender for a while, which weighed 47tons 6cwt — four tons heavier than the locomotive. After withdrawal of the RODs, these tenders were first fitted to the 'Aberdares' and then to Colletts 0-6-0s Nos 2281-86 on construction, but other 2251 class engines ran with them for brief periods. *J. Spencer Gilks*

75006

2255

TALERDDIG
6368

7828

Above left:
Preparing to exchange tokens at Talerddig, ex-GWR 4300 class Mogul No 6368 enters the station slowly with an early morning train on 29 August 1963. *Andrew Muckley*

Left:
On 29 August 1963, a westbound train, headed by GWR 'Manor' No 7828 *Odney Manor* is seen at Llanbrynmair. The small signal cabin is just visible on the left. *Andrew Muckley*

Above:
'Manor' No 7828 *Odney Manor* passes Commins Coch Halt with the 12.35pm Aberystwyth to Shrewsbury train on 12 September 1964. At this point, the main A489 trunk road from Newtown to Machynlleth runs parallel with the track for some considerable distance. *J. C. Beckett*

Cemmes Road

Above:
A mixed train, with a goods van between the first and second coaches, stops at Cemmes Road station to receive attention to its locomotive, 4-4-0 No 64, in July 1909. The train is the 10.55am freight from Aberystwyth. No 64 eventually became GWR No 1093 and was withdrawn in April 1926, after 32 years' service. No 64 ran with large rectangular numberplates for a while. *H. L. Hopwood*

Below:
With an Aberystwyth to Oswestry train, GWR 'Manor' No 7822 *Foxcote Manor* steams into Cemmes Road station in August 1964. Note the facing signal situated at the end of the platform, beside the up line. *Andrew Muckley*

Machynlleth

Above:
Collett 0-6-0 No 2268 heads an evening Shrewsbury to Aberystwyth working and is seen at Machynlleth in August 1964. By this time, less than 25% of the class remained in service. *Andrew Muckley*

Below:
Passing the goods shed, on entering Machynlleth station, the 3.45pm Shrewsbury to Aberystwyth train, hauled by 'Dukedog' No 9017, coasts in to set down and pick up passengers. The platforms here were staggered, the down platform passing over the main road. *T. K. Widd*

11:

The Dinas Mawddwy Branch

The privately-owned Mawddwy Railway, built mainly for slate traffic, opened in 1867, and ran from Cemmes Road, on the main line — a distance of 6¾ miles. The slate was conveyed to the port of Aberdovey. Passenger trains also ran from the opening of the line, but it closed, under private ownership on 17 April 1901 to passengers, and in April 1908 to freight.

However, it was reopened by the Cambrian Railways on 31 July 1911 and remained open to passenger traffic until January 1931, and to freight until 1950.

After the Cambrian reopening, Sharp Stewart 2-4-0s operated mixed trains and school trains, which had previously only operated to Cemmes Road, where the children changed trains. School trains now ran through to Machynlleth.

The original locomotives which operated the line were two Manning Wardle 0-6-0STs — *Mawddwy* and *Disraeli*. *Mawddwy* was originally named *Alyn* and was involved in the construction of the Potteries, Shrewsbury & North Wales Railway. *Mawddwy* was reconditioned at Oswestry in late-1910 for the line's reopening — *Disraeli* was broken up in 1912. In 1913 *Mawddwy* was numbered 30 (eventually GWR No 824) — it was withdrawn in September 1940. Its last days were spent on the Tanat Valley Light Railway and the Van branch.

In 1923, three passenger trains per day ran in each direction, and there was no Sunday service, and in May 1927, permission was granted to use six-wheeled coaching stock over the branch.

By 1947, as the line's life was nearing its end, on Mondays, Wednesdays and Fridays a freight train ran to and from Machynlleth, taking about 1hr 10min to complete the 6 mile 63 chain journey — double-heading was allowed if necessary. Speed could not exceed 25mph, and over reverse curves was not allowed to rise above 15mph in either direction. At this time, locomotives permitted to operate these freight trains were 2-4-0T No 1308, 0-6-0PTs (2021 class), and ex-Cambrian engines Nos 819, 1196, 1197, 898, 900 and 908.

At the end, GWR Collett 0-4-2 tanks (1400 class) were operating the short freights on the branch, having been on the line when in their 4800 state, prior to their renumbering.

Along the route, at Aberangell, various narrow-gauge lines, known as the Hendre-ddu Tramways, originally joined the Mawddwy Railway and conveyed slate, making Aberangell an exchange station for this type of merchandise.

Closure of the line was caused by the River Dovey weakening the railway bridge situated on the Aberangell side of Cemmes station, where the line was inclined at 1 in 83.

Today, Dinas Mawddwy station building still stands, as a café. There is also a Welsh woollen mill on the site. However, on visiting this location today, one still gets the atmosphere of a branch line terminus, with the old buildings and water column still in evidence.

Below:
Aberangell station was midway along the 6¾-mile branch from Cemmes Road to Dinas Mawddwy and, in September 1949, 'Dean Goods' No 2323 is seen with a freight, conveying timber stakes in the first wagon. *Oakwood Collection*

Above:
On 24 September 1948, 0-4-2T No 1434 stands in Dinas Mawddwy station with a short freight. Goods trains continued to operate until 1950 and the station buildings now house a café. *R. S. Carpenter Collection*

Below:
Signs of the goods shed and water column at Dinas Mawddwy are still apparent today, and on 24 September 1948, 0-4-2T No 1434 is seen taking water at this overgrown location.
R. S. Carpenter Collection

12:

Machynlleth to Aberystwyth

The line from Machynlleth to Aberystwyth was opened in two stages – to Borth on 1 July 1863, and from there to Aberystwyth on 23 June 1864 – this was part of the Aberystwyth & Welsh Coast Railway. The section from Machynlleth towards Aberystwyth of the A&WCR was the first to be worked upon, as the Newtown & Machynlleth Railway had succeeded in getting a clause placed within the Bill, incorporating the A&WCR, which stated that if they were unable to construct the line from Aberystwyth to Machynlleth by 1 August 1864, then the N&M could take over the job of completing the link – extending their own railway, as Aberystwyth was a major resort.

As the line travelled west, Dovey Junction was reached – it was formerly called Glandovey Junction. The name was changed to avoid confusion with the next station on the line, Glandovey, which was made Glandyfi at the same time. Dovey Junction station had no exit as passengers only normally changed trains here. Before the change in county names, the station area was, in fact, situated in three different counties – Merioneth (the stationmaster's house), Montgomery (the station), and Cardigan (the up distant signal). The station was sited within marshland.

Many trains from Pwllheli terminated at Dovey Junction where connecting services to Aberystwyth and Machynlleth were laid on. The station only had two platform faces making it difficult at times when the detaching of portions of trains was necessary. At this location Pwllheli portions of the 'Cambrian Coast Express' would be detached/attached and the 'Manors' were used generally on Aberystwyth portions only.

In 1958, the old station buildings were replaced by modern structures and the old signalbox closed, a new one appearing on the station platform – 52 levers were replaced by 65.

After Glandyfi, Ynys Las was reached, from where a ferry plied to Aberdovey after the line to Borth had been completed. However it was originally proposed to build a bridge from Ynys Las to Aberdovey, and work on this project actually commenced, but this was abandoned, and the present route via Dovey Junction was chosen after various problems concerning boring occurred – men could only work at low tide and, at high tide, leisure hours were spent in adjacent hostels. After that, the men were not really 'fit for work'. But, in all fairness, a firm foundation in the sand was difficult to find for the bridge to be built.

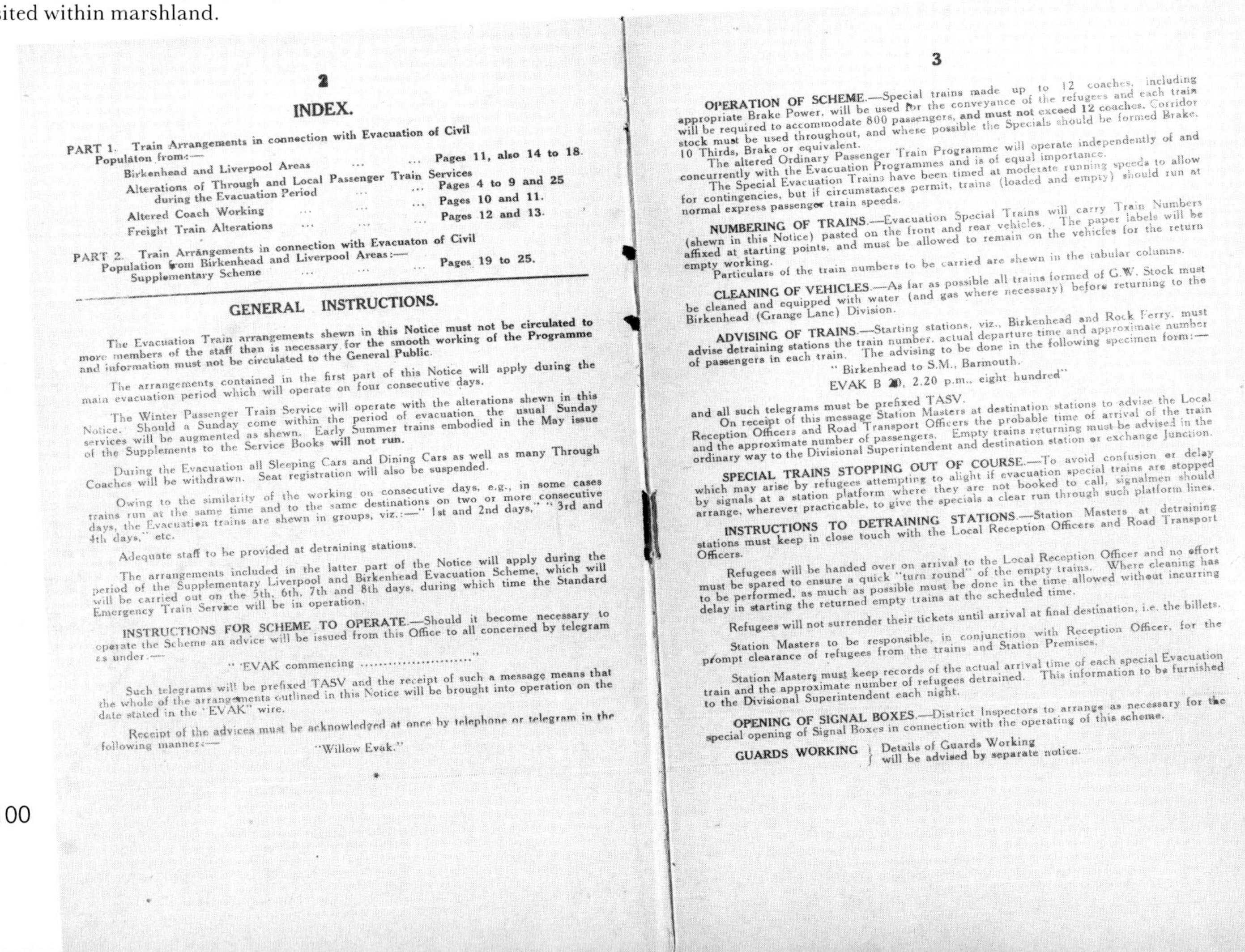

2

INDEX.

PART 1. Train Arrangements in connection with Evacuation of Civil Populaton from:—
Birkenhead and Liverpool Areas ... ... Pages 11, also 14 to 18.
Alterations of Through and Local Passenger Train Services during the Evacuation Period ... ... Pages 4 to 9 and 25
Altered Coach Working ... ... Pages 10 and 11.
Freight Train Alterations ... ... Pages 12 and 13.

PART 2. Train Arrangements in connection with Evacuaton of Civil Population from Birkenhead and Liverpool Areas:— Supplementary Scheme ... ... Pages 19 to 25.

GENERAL INSTRUCTIONS.

The Evacuation Train arrangements shewn in this Notice must not be circulated to more members of the staff than is necessary for the smooth working of the Programme and information must not be circulated to the General Public.

The arrangements contained in the first part of this Notice will apply during the main evacuation period which will operate on four consecutive days.

The Winter Passenger Train Service will operate with the alterations shewn in this Notice. Should a Sunday come within the period of evacuation the usual Sunday services will be augmented as shewn. Early Summer trains embodied in the May issue of the Supplements to the Service Books **will not run.**

During the Evacuation all Sleeping Cars and Dining Cars as well as many Through Coaches will be withdrawn. Seat registration will also be suspended.

Owing to the similarity of the working on consecutive days, e.g., in some cases trains run at the same time and to the same destinations on two or more consecutive days, the Evacuation trains are shewn in groups, viz.:—" 1st and 2nd days," " 3rd and 4th days," etc.

Adequate staff to be provided at detraining stations.

The arrangements included in the latter part of the Notice will apply during the period of the Supplementary Liverpool and Birkenhead Evacuation Scheme, which will will be carried out on the 5th, 6th, 7th and 8th days, during which time the Standard Emergency Train Service will be in operation.

INSTRUCTIONS FOR SCHEME TO OPERATE.—Should it become necessary to operate the Scheme an advice will be issued from this Office to all concerned by telegram as under:—

" 'EVAK commencing ...."

Such telegrams will be prefixed TASV and the receipt of such a message means that the whole of the arrangements outlined in this Notice will be brought into operation on the date stated in the "EVAK" wire.

Receipt of the advices must be acknowledged at once by telephone or telegram in the following manner:— "Willow Evak."

3

OPERATION OF SCHEME.—Special trains made up to 12 coaches, including appropriate Brake Power, will be used for the conveyance of the refugees and each train will be required to accommodate 800 passengers, **and must not exceed 12 coaches.** Corridor stock must be used throughout, and where possible the Specials should be formed Brake, 10 Thirds, Brake or equivalent.

The altered Ordinary Passenger Train Programme will operate independently of and concurrently with the Evacuation Programmes and is of equal importance.

The Special Evacuation Trains have been timed at moderate running speeds to allow for contingencies, but if circumstances permit, trains (loaded and empty) should run at normal express passenger train speeds.

NUMBERING OF TRAINS.—Evacuation Special Trains will carry Train Numbers (shewn in this Notice) pasted on the front and rear vehicles. The paper labels will be affixed at starting points, and must be allowed to remain on the vehicles for the return empty working.

Particulars of the train numbers to be carried are shewn in the tabular columns.

CLEANING OF VEHICLES.—As far as possible all trains formed of G.W. Stock must be cleaned and equipped with water (and gas where necessary) before returning to the Birkenhead (Grange Lane) Division.

ADVISING OF TRAINS.—Starting stations, viz., Birkenhead and Rock Ferry, must advise detraining stations the train number, actual departure time and approximate number of passengers in each train. The advising to be done in the following specimen form:—

" Birkenhead to S.M., Barmouth.
EVAK B 20, 2.20 p.m., eight hundred"

and all such telegrams must be prefixed TASV.

On receipt of this message Station Masters at destination stations to advise the Local Reception Officers and Road Transport Officers the probable time of arrival of the train and the approximate number of passengers. Empty trains returning must be advised in the ordinary way to the Divisional Superintendent and destination station or exchange Junction.

SPECIAL TRAINS STOPPING OUT OF COURSE.—To avoid confusion or delay which may arise by refugees attempting to alight if evacuation special trains are stopped by signals at a station platform where they are not booked to call, signalmen should arrange, wherever practicable, to give the specials a clear run through such platform lines.

INSTRUCTIONS TO DETRAINING STATIONS.—Station Masters at detraining stations must keep in close touch with the Local Reception Officers and Road Transport Officers.

Refugees will be handed over on arrival to the Local Reception Officer and no effort must be spared to ensure a quick "turn round" of the empty trains. Where cleaning has to be performed, as much as possible must be done in the time allowed without incurring delay in starting the returned empty trains at the scheduled time.

Refugees will not surrender their tickets until arrival at final destination, i.e. the billets.

Station Masters to be responsible, in conjunction with Reception Officer, for the prompt clearance of refugees from the trains and Station Premises.

Station Masters must keep records of the actual arrival time of each special Evacuation train and the approximate number of refugees detrained. This information to be furnished to the Divisional Superintendent each night.

OPENING OF SIGNAL BOXES.—District Inspectors to arrange as necessary for the special opening of Signal Boxes in connection with the operating of this scheme.

GUARDS WORKING } Details of Guards Working will be advised by separate notice.

The first trains to leave Borth for Machynlleth were packed to capacity by enthusiastic travellers. A hundred filled the first train and, only 4½hr later, a further 530 fought for seats on the second train. The trains left through an avenue of appreciative men and women who had travelled from Aberystwyth. Hymns were sung and a good time was had by all. The journey took about 25min – a distance of around 12 miles.

By 1872, a refreshment room was opened at Borth where first-class and third-class passengers were segregated. Until the late-1880s, refreshment rooms on the Cambrian were run by local innkeepers.

As the line left the coast and headed south towards Aberystwyth, Llandre was reached. In November 1952, a fierce gale brought in a 17ft tide at Aberystwyth and Borth. Salt-coated rails hampered GWR 'Manor' No 7803 *Barcote Manor* in making reasonable progress, having taken 12min to cover the last 2 miles into Borth from Ynys Las. Owing to an obstruction near Llandre, resulting from the gale, the train was held at Borth. High winds and flooding had caused many problems around the Dovey Junction and Aberystwyth areas, and in December 1962, flooding caused the trackbed to be washed away on the Derwenlas curve, two miles east of Dovey Junction, leaving 200yd of track suspended in the air. Complete repair to the track was completed in just nine days.

The last station before Aberystwyth was reached was Bow Street where the main station buildings were on the down platform, whereas at Llandre, they were on the up side – at Borth, they were also on the up platform. From Borth to Aberystwyth gradients did not exceed 1 in 75.

As the line approached Aberystwyth, the main line ran parallel with the Vale of Rheidol Railway from Llanbadarn, just over a mile from the terminus. The Great Western route from Carmarthen (closed 1964/65) crossed the Vale of Rheidol line and entered the station by passing behind the standard gauge engine shed within yards of the turntable (its later position).

Improvements to the A&WCR station at Aberystwyth were not made until 1924, by the GWR, who constructed a two-storey building of stone and granite. These alterations included the extension of Platforms 3 and 4 to 856ft and Platform 5 (for Carmarthen services) to 605ft.

At the engine shed, opened in 1891, the turntable was moved south of the running shed, which proved necessary when the platforms were lengthened. This was a two-road shed, and in 1938 the engine shed was replaced by a new building on the same site, and extra buildings and a lifting gantry were installed. A 45,000gal water tank was also located here.

When the ex-Metropolitan 4-4-0Ts were purchased by the Cambrian, Nos 12, 33, 36 and 37 were sent from Oswestry, where their condensing gear had been removed, to Aberystwyth for local passenger work. In fact, from July 1915, No 12 worked as far as Barmouth.

Aberystwyth shed (ABH) had its own coded-allocation of locomotives until 1932, after which it became a sub-shed of Machynlleth (MCH/89C).

Four-coupled Metro tank No 617 (built in 1871 at Swindon) was allocated to Aberystwyth in 1931, and was the only one of its class left on the Central Wales Division. The 1931 allocation also included 'Dukes'. By 1938, eight 'Earls' were sub-shedded at Aberystwyth and, after 1939, GWR 4300 class Moguls were based there. Sub-shedded at this location in December 1947 were 'Manors' Nos 7802/03; 'Dukedogs' Nos 9002/13/17/21/25; 'Duke' No 9087; Collett 0-6-0s Nos 2200/23/60/83/98; 2-6-2Ts Nos 5560/70; and 0-6-0PT No 7406. 'Dukedogs' in store there in August 1956 included Nos 9016 and 9022.

By 1923, 'Cambrian' daily arrivals at Aberystwyth station totalled 13 (but only one on Sundays), and 'GWR' services from Carmarthen stood at six (weekdays only), one of these trains only operating from Lampeter.

According to a 'Diversion Notice' of December 1943, obstructions between Dovey Junction and Aberystwyth could result in trains being diverted via Carmarthen – a detour that does not bear thinking about!

During World War II, 'Evacuation Specials' conveyed passengers from Overton-on-Dee to Borth, and from Bettisfield to Aberystwyth, laid on for the population of Liverpool and Birkenhead. On the eve of the first day of evacuation, empty coaching stock travelled from Aberystwyth to Gobowen, each train made up of 10 coaches and two brakes and, wherever possible, spare corridor thirds replaced composites and non-corridor thirds. The accompanying illustrations are reproduced from *GWR Notice No 283* which was prepared for an eight-day programme of evacuation from London, Liverpool and Birkenhead. The notice is undated but is very likely to have been prepared for the first wave of civilian evacuees which took place at the beginning of September 1939.

4

Suspension of Passenger Trains.

On the 1st, 2nd, 3rd and 4th days of the Evacuation period the undermentioned Trains

WILL NOT RUN.

Whitchurch, Shrewsbury and Aberystwyth.

DOWN TRAINS. WEEK-DAYS.

8 20 a.m. Oswestry to Aberystwyth.
9 35 a.m. (MO) Oswestry to Welshpool.
1 0 p.m. (WSO) Oswestry to Llanfyllin.
1 52 p.m. Whitchurch to Aberystwyth.
2 20 p.m. Newtown to Moat Lane.
(SO and last Tuesday).
3 24 p.m. (SO) Machynlleth to Aberystwyth.
3 5 p.m. Shrewsbury to Welshpool.
6 35 p.m. Machynlleth to Aberystwyth.
9 0 p.m. Whitchurch to Oswestry.

UP TRAINS. WEEK-DAYS.

10 15 a.m. Aberystwyth to Whitchurch.
11 55 a.m. Llanfyllin to Oswestry.
(WSO)
2 10 p.m. Welshpool to Oswestry.
(MO)
1 40 p.m. (ety) Moat Lane to Newtown.
(SO and last Tuesday)
12 5 p.m. Aberystwyth to Shrewsbury.
5 5 p.m. Aberystwyth to Machynlleth.
5 35 p.m. Welshpool to Whitchurch.
7 30 p.m. Aberystwyth to Machynlleth.
9 10 p.m. Aberystwyth to Machynlleth. (SO)

For alterations of other services and additional trains, see pages 6 and 7.

TANAT VALLEY BRANCH.

8 45 p.m. Oswestry to Llangynog.
(1st Wednesday.)
10†10 p.m. Llangynog to Oswestry.
(No Sunday Service.)

LLANFYLLIN BRANCH.

11 55 a.m. Llanfyllin to Oswestry.
(WSO)
1 0 p.m. Oswestry to Llanfyllin.
(WS.O)
(No Sunday Service.)

MOAT LANE AND BRECON.

3 55 p.m. Moat Lane to Llanidloes.
10 15 a.m. Llanidloes to Brecon.
(For Sunday Service, see pages 8 and 9).

WEEK-DAYS.

1 20 p.m. Brecon to Moat Lane.
4 20 p.m. Llanidloes to Moat Lane.

MACHYNLLETH, RUABON AND PWLLHELI. WEEK-DAYS.

11 5 a.m. Dovey Jct. to Barmouth.
12 0 noon Portmadoc to Pwllheli.
3 5 p.m. Ruabon to Pwllheli. (SO)
3 10 p.m. Ruabon to Barmouth. (SO)
3 15 p.m. Dolgelley to Barmouth.
4 8 p.m. Barmouth to Pwllheli.
4 38 p.m. Dovey Jc. to Barmouth.

9 35 a.m. Pwllheli to Ruabon. (MO)
9 40 a.m. Barmouth to Dovey Jct.
10 15 a.m. Barmouth to Ruabon (MO)
2 23 p.m. Barmouth to Dovey Jct.
3 30 p.m. Pwllheli to Portmadoc.
7 16 p.m. Portmadoc to Dolgelley.

Alterations of Other Services.

9 5 a.m. Pwllheli to Ruabon (M.O.)
3 5 p.m. Ruabon to Pwllheli (S.X.)
8 20 p.m. Dovey Jct. to Portmadoc
8 55 p.m. Portmadoc to Machynlleth
Will run daily.
(For Sunday Service, see pages 8 and 9).

WREXHAM AND ELLESMERE BRANCH
(No alterations and no Sunday Services.)

OSWESTRY AND GOBOWEN BRANCH.
(See page 5).

BRECON AND MERTHYR

WEEK DAYS.

6 45 p.m. Newport to Brecon will start at 7.5 p.m. and run 20 minutes later throughout.

SUNDAY.

The undermentioned week-day trains will run on Sunday:—
2 0 p.m. Brecon to Newport.
6 10 p.m. Brecon to Newport.
2 50 p.m. Newport to Brecon.
6 45 p.m. Newport to Brecon starting at 7.5 p.m. and running 20 minutes later throughout.

CARMARTHEN, ABERYSTWYTH AND ABERAYRON BRANCH.

4 35 p.m. Aberayron to Felin Fach.
5 0 p.m. Felin Fach to Aberayron.

Alterations of Other Services.

2 10 p.m. Carmarthen will be re-timed. For details see Swansea Division Notice No. T3297.
3 40 p.m. Lampeter to Aberayron to be held at Lampeter to make connection and to work forward from Felin Fach in times of School train.

Far left:
The General Instructions issued to the Central Wales Division for the civilian evacuation from London, Liverpool and Birkenhead.

Left:
Trains to be suspended during the first four days of the evacuation.

Bottom:
With the Barmouth portion of the 'Cambrian Coast Express', 4500 class 2-6-2T No 5510, recently transferred from Swindon and withdrawn by 1960, stands in Machynlleth station in 1959 sporting the official headboard. The stone-block station buildings are clearly visible.
Norman Preedy

Far right:
This view of Machynlleth station clearly shows the basic timber building on the up platform. 'Manor' No 7828 *Odney Manor* is seen entering the station, in August 1964, with an Aberystwyth to Oswestry train.
Andrew Muckley

Bottom right:
Differing motive power stands outside Machynlleth engine shed in August 1959. 'Dukedog' No 9017 and 2-6-2T No 5556 prepare for their next duty. At this time, six 'Dukedogs' remained in service, and three of these were allocated to Croes Newydd (84J).
Norman Preedy

Below:
Evacuation specials to Aberystwyth on the 5th, 6th, 7th and 8th days of the evacuation.

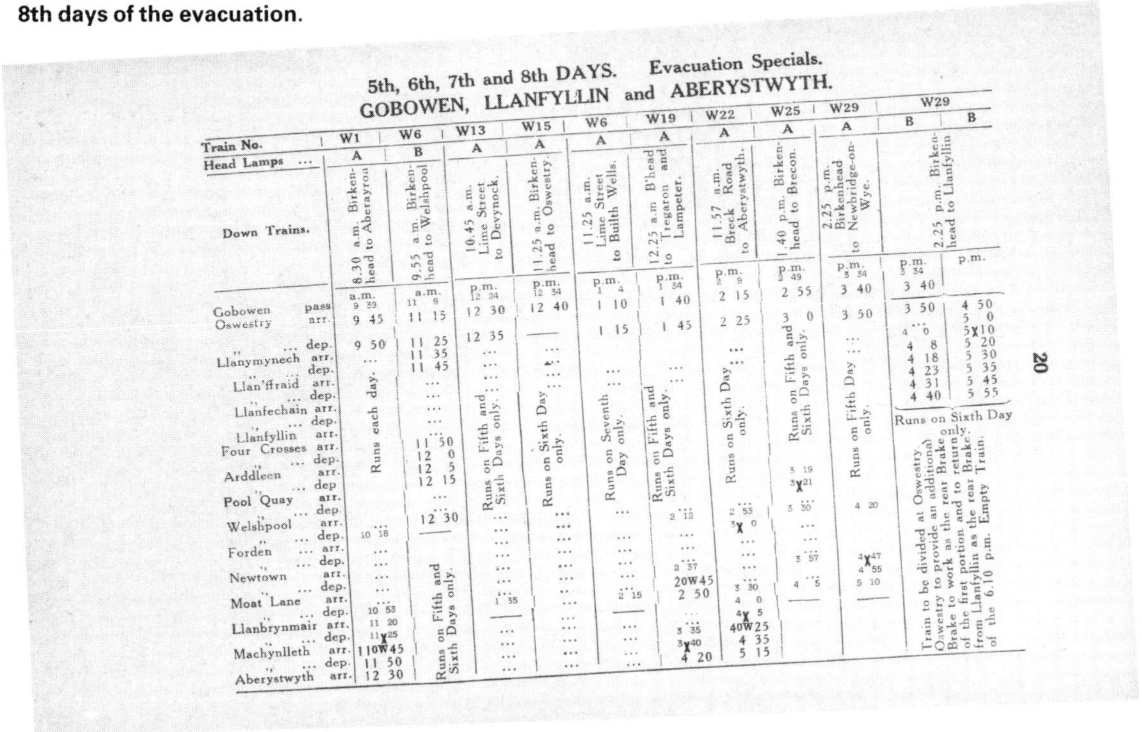

5th, 6th, 7th and 8th DAYS. Evacuation Specials.
GOBOWEN, LLANFYLLIN and ABERYSTWYTH.

Train No.	W1	W6	W13	W15	W6	W19	W22	W25	W29	W29	
Head Lamps ...	A	B	A	A	A	A	A	A	A	B	B
Down Trains.	8.30 a.m. Birkenhead to Aberayron	9.55 a.m. Birkenhead to Welshpool	10.45 a.m. Lime Street to Devynock.	11.25 a.m. Birkenhead to Oswestry.	11.25 a.m. Lime Street to Builth Wells.	12.25 a.m B'head to Tregaron and Lampeter.	11.57 a.m. Breck Road to Aberystwyth.	1.40 p.m. Birkenhead to Brecon.	2.25 p.m. Birkenhead to Newbridge-on-Wye.	2.25 p.m. Birkenhead to Llanfyllin	
	a.m.	a.m.	p.m.	p.m.	p.m.	p.m.	p.m.	p.m.	p.m.	p.m.	p.m.
Gobowen pass	9 39	11 9	12 24	12 34	1 4	1 34	2 9	2 49	3 34	3 34	
Oswestry arr.	9 45	11 15	12 30	12 40	1 10	1 40	2 15	2 55	3 40	3 40	
,, ... dep.	9 50	11 25	12 35	—	1 15	1 45	2 25	3 0	3 50	3 50	4 50
Llanymynech arr.	...	11 35	...	...			...		...	4 0	5 0
,, ... dep.		11 45	...	...	...	...	...		...	4 8	5X10
Llan'ffraid arr.		...	...	...	...	...				4 18	5 20
,, ... dep.		...	...							4 23	5 30
Llanfechain arr.		...								4 31	5 35
,, ... dep.		...									5 45
Llanfyllin arr.		...								4 40	5 55
Four Crosses arr.		11 50									
,, ... dep.		12 0									
Arddleen arr.		12 5						3 19			
,, ... dep		12 15						3X21			
Pool Quay arr.		...					...	...			
,, ... dep.		...			...	2 12	2 53	3 30	4 20		
Welshpool arr.	...	12 30	...	...			3X0	...			
,, ... dep.	10 18		...	...	...	...	...	...			
Forden ... arr.	...		...	...	...	...	...	...	4X47		
,, ... dep.	...		...	...	...	...	...	3 57			
Newtown arr.	...		...	...	...	2 37	...	...	4 55		
,, ... dep.	...		...	...	...	20W45	3 30	4 5	5 10		
Moat Lane arr.	...		1 35	...	2 15	2 50	4 0				
,, ... dep.	10 53			...		...	4X5				
Llanbrynmair arr.	11 20		...	...	...	...	40W25				
,, ... dep.	11X25		...	...	...	3 35	4 35				
Machynlleth arr.	1 10W45		...	...	...	3X40	5 15				
,, ... dep.	11 50		...	...	...	4 20					
Aberystwyth arr.	12 30										
	Runs each day.	Runs on Fifth and Sixth Days only.	Runs on Fifth and Sixth Days only.	Runs on Sixth Day only.	Runs on Seventh Day only.	Runs on Fifth and Sixth Days only.	Runs on Sixth Day only.	Runs on Fifth and Sixth Days only.	Runs on Fifth Day only.	Runs on Sixth Day only. Train to be divided at Oswestry. Oswestry to provide an additional Brake to work as the rear Brake of the first portion and to return Brake from Llanfyllin as the rear Brake of the 6.10 p.m. Empty Train.	

20

7828

9017

Right:
Machynlleth engine shed, with BR Standard 2-6-4T No 80099 sat on the short road which terminated within the depot building. This engine was one of the few members of this class to be allocated here from around 1962. On this occasion it shares the depot with No 7803 *Barcote Manor*, whilst Class 3 BR 2-6-2T No 82021 prepares to drop its fire. *Andrew Muckley*

Right:
Single-handed operation at Machynlleth's turntable, situated beneath the sheer rock face at this location. A 2-6-2T, No 4575, is turned in August 1951.
Oakwood Collection

Below:
To move this Standard 4-6-0, No 75002, it required two men to propel the hand-operated turntable at Machynlleth. The scene was captured in August 1964.
Andrew Muckley

Dovey Junction

Left:
Dovey Junction, before new station buildings and a platform-sited signalbox came into operation. On 20 July 1948, 'Dukedog' No 9072 prepares to back on to the 1.35pm train to Pwllheli. *Oakwood Collection*

Left:
Great Western 'Duke' No 3263 *St Michael* is pictured at Dovey Junction with an interesting four-coach train. Originally numbered 3267, this locomotive, built in 1896, was withdrawn in March 1936.
Author's Collection

Below:
The rebuilt Dovey Junction station on 13 August 1959, as 2-6-2T No 5556 enters with a local service. The new-style buildings, lamps and station sign are clearly visible.
R. S. Carpenter Collection

Above:
Borth station, with its clerestory camping coach in the old bay platform — in use (so it seems) at the time. Ivatt 2-6-0 No 46518 stops with its train to pick up and set down passengers whilst en route to Shrewsbury. *Andrew Muckley*

Below:
Bow Street station on 29 August 1963. No 7822 *Foxcote Manor* is pictured with a train bound for Aberystwyth. Note the weeds which are gradually taking over. *Andrew Muckley*

Aberystwyth

Above:
GWR 'Manor' No 7810 *Draycott Manor* awaits departure from Aberystwyth, on 23 August 1963, with a Whitchurch and Shrewsbury train. By early 1965, No 7810 had been withdrawn. *R. N. Joanes*

Below:
The 9.55am Shrewsbury service prepares to leave Aberystwyth, on 14 August 1958, behind 4300 class Mogul No 6378. The platform awning, covering the entire platform width, is clearly visible. *Brian Morrison*

'Deans' on shed

Left:
Aberystwyth shed, and 4-4-0 'Duke' No 3256 *Guinevere*, in immaculate condition, stands on the coaling line. It sports its original cab with circular windows and its whistle on the cab roof. It is also fitted with the strengthened frame. *Real Photographs/5373*

Below left:
GWR 'Duke' No 3287 *Mercury* (originally No 3321) remained in service until July 1949, and in 1930 was fitted with a 3,500gal Churchward tender. This created problems concerning draughts, etc, as the tender was wider than the cab. On certain 'Dukes' fitted with these wider tenders, cab side-sheets were set outwards at the rear to create a smooth air-flow, but No 3287 was not modified in this way. Renumbered as 9087, with an unusually-placed numberplate, it is seen at Aberystwyth shed on 20 July 1948 after breaking a rear axle. Note the newer-style cab compared with the previous picture.
Oakwood Collection

Bottom left:
Two 4-4-0 'Dukedogs' stand at Aberystwyth shed. Basically, these engines were rebuilt 'Bulldogs' as their frames were used, but they also took their cabs from scrapped 'Dukes'. No 9022, seen here, had the frame of 'Bulldog' No 3436 and 'officially replaced' the 'Duke' No 3278 *Trefusis*.
H. N. James

13:

The Vale of Rheidol Railway

By the 1890s, Aberystwyth was well established as a popular seaside resort, and Devil's Bridge was a well-known beauty spot — lead was also mined in the Rheidol Valley. The construction of the 1ft 11½in-gauge line reached Devil's Bridge on 28 July 1902, but excessively sharp curves resulted in a Board of Trade refusal for passenger traffic. However, freight traffic commenced operations in August 1902, and exceeded all expectations, and on 22 December of that year, passenger services started, after additional work enabled the line to be passed by the Board of Trade.

In addition to lead ore traffic, which converged on Devil's Bridge station from the many mines around, timber traffic also developed.

On 7 March 1904, Rheidol Falls Halt was added, and in May, Glanrafon Halt came into being. Rhiwfron Halt, built around 1905, was the final station on the line to be opened, and at the same time a proper station was constructed at Aberffrwd, with a passing loop, and watering facilities, replacing the inconveniently-placed facilities at Nantyronen.

Freight traffic declined in the mid-1920s and operations of this nature ceased. From 1 January 1931, winter passenger services also discontinued, making one locomotive surplus to requirements — the axe fell on No 1212 (ex-No 1) as it was the engine in the worst condition. It was sent to Swindon in 1932.

New coaching stock was purchased in 1938, specially designed to attract more tourists, but the line closed in 1939 as war broke out. Maintenance to permanent way and stock continued during the war, and a limited service restarted on 23 July 1945.

After the 1954 summer season, the line was threatened with closure but, as the tourist potential was so great, a mammoth advertising campaign produced the highest passenger figures to date, in 1954 — 23,616 (a 7,000 increase). By 1963 passenger figures had reached 34,237.

In 1989, BR sold the line to the Brecon Mountain Railway who operated services over this picturesque line.

At Aberystwyth, in 1924, the Vale of Rheidol Railway station was moved closer to the main line station and, from 1968, the VoR operated from its platforms. The VoR engine shed was situated on the opposite side of Station Road to the station, near the Crosville bus garage. The shed area comprised a corrugated-iron-clad timber-framed two-road shed, with entrances staggered, giving undercover shelter of 80ft on one side, and 45ft on the other. In 1925, the GWR installed a water tank. The area also included a coaling bay and offices. From 1968, VoR locomotives were accommodated in the former standard-gauge shed.

Following the route to Devil's Bridge, the first station was Llanbadarn, where there were booking and waiting facilities and a goods siding. Capel Bangor was next, which originally was to have been named Rhiwarthen after a nearby farm.

Below:
The Davies & Metcalfe 2-6-2T, No 1 *Edward VII*, was renumbered 1212 in 1923 and cut up in March 1935 after being sent to Swindon following withdrawal as surplus to requirements in 1932. *Real Photographs*

Originally this was the main intermediate station, and a carriage shed and passing loop were situated there. Before the line opened, Nanteronen station was scheduled to be named Tyllwyd — then Rhydyronen — and its siding faced Devil's Bridge. The Rheidol mine used Rhiwfron Halt for loading purposes by the use of an aerial ropeway. All other stations were added after 1904, and are mentioned earlier.

At Devil's Bridge, the original goods yard lost its loop and continued as two separate sidings. Watering facilities existed and the station is still a hive of activity today as trains arrive from Aberystwyth. The entire line covers a distance of about 11½ miles, the steepest gradient, between Aberffrwd and Devil's Bridge being 1 in 50, over 3¼ miles.

The first Vale of Rheidol engines to operate the line were built by Davies & Metcalfe of Manchester — No 1 *Edward VII* and No 2 *Prince of Wales*, both 2-6-2Ts. No 3 *Rheidol*, a 2-4-0T, was added to VoR stock in 1903. Originally, the 2-6-2Ts were painted green with panelling of double-white lines, edged with black. The company crest appeared on the tank-side, with the nameplate. Later, the nameplate was removed and the word 'CAMBRIAN' appeared — this was after the engines has been repainted in Cambrian colours. Various other changes appeared, including number changes to No 2 (1213 and 9 resulting in a totally new locomotive appearing in 1924), and in 1923, two Swindon-built 2-6-2Ts were added to the ranks, Nos 7 and 8, about the same time as No 3 *Rheidol* (GWR No 1198) ended its days.

Today, *Owain Glyndŵr*, *Llywelyn* and *Prince of Wales* are the pride and joy of the Vale of Rheidol — who knows what livery changes will appear under the new ownership.

Prince of Wales

Left:
No 3 *Rheidol* was a 2-4-0T, built by Bagnall of Stafford in 1896. It is pictured at Aberystwyth in 1903 with the up mail. A telegraph pole is seen carried on the carriage step. This locomotive was originally built for working coal mines in Brazil on a 750mm-gauge track. However, it never reached South America. *Ian Allan Library*

Below left:
Another view of No 3 *Rheidol*, at Aberystwyth on 28 June 1909. The spark arresting chimney, seen in the previous view, was modified in 1904, and the number now appears on the modified buffer beam. No 3 became GWR No 1198, was always painted black and was withdrawn in 1923.
LCGB/Ken Nunn Collection

Above:
No 9 *Prince of Wales* passes the Crosville bus garage at Aberystwyth on 4 June 1963 with a train for Devil's Bridge. After nationalisation, No 9 was renumbered from 1213 and was painted black, but soon was to be painted in main-line green. After a period without its nameplate, it was refitted in 1956. *T. J. Edgington*

Below:
Scampering across the road between the narrow gauge shed and Aberystwyth station is No 9 *Prince of Wales*, on 27 August 1960, about to haul the 10am train to Devil's Bridge. No 9 was given the 'yellow ochre' livery in 1981. *J. C. Haydon*

Awaiting Departure

Top:
At the height of the tourist season, on 1 August 1960 — August Bank Holiday Monday — No 7 *Owain Glyndŵr* prepares to depart from Aberystwyth with the 1.45pm service to Devil's Bridge. No 9 *Prince of Wales* stands alongside with the 2.30pm working. *M. Mensing*

Above:
Awaiting departure from its 'new' terminus — Aberystwyth's main line station — No 7 *Owain Glyndŵr* is seen with the 1.40pm relief on the VoR Open Day, on 4 May 1981. Note the raised BR 'Arrow' emblem now fitted to the cab-side. From 1981, No 7 was lined out in BR blue. *British Rail*

Devil's Bridge

Above right:
Taking on water at Devil's Bridge is No 8 *Llywelyn*, with the BR double-arrow emblem of the tankside. It has just worked the 10.15am service from Aberystwyth, on 2 June 1977. It carries the plain BR blue livery. *G. B. Wise*

Right:
Devil's Bridge station, and No 8 *Llywelyn*, in BR blue livery, awaits departure to Aberystwyth with its four-coach train on the afternoon of 4 October 1973. No 8, in 1981, was painted in GWR green livery. *Brian Morrison*

14:

Dovey Junction to Pwllheli

This was the major part of the Aberystwyth & Welsh Coast Railway's line, and covered a distance of about 54 miles. From Dovey Junction, the route followed a comparatively level terrain to Aberdovey, passing through four tunnels as the line skirted the banks of the Dovey estuary. These were Frongoch (199¾yd), Morfor (219yd), Penhellyg (119yd) and Craig-y-Don (533yd), and straight sections of track along this route were unheard of. The line from Dovey Junction to Aberdovey opened to traffic on 14 August 1867 and although, originally, no stations appeared on this stretch of line, the GWR installed three halts, at Gogarth, Abertafol and Penhelig. The first two halts were unstaffed and comprised single wooden platforms with a centrally-placed waiting shelter. Penhelig's platform was much longer than those at Abertafol and Gogarth (75ft and 99ft respectively), measured 200ft, and sported a 'Pagoda' design shelter. It was sandwiched between Penhellyg and Craig-y-Don tunnels.

Aberdovey was an important seaport in its early days, and from the station a branch line ran down to the harbour where an interesting wagon turntable was situated. This branch ceased operations in 1964, In 1904, an announcement was made, within the Cambrian timetable, advertising excellent facilities at Aberdovey's new wharf and a deep-water pier for freight services to Southern Ireland, Rosslare harbour being only 96 miles away. There were direct rail links to Aberdovey from virtually everywhere in Britain and the advertisement showed mileages from towns suitably situated to use the port. Aberdovey had, since the 1880s, been operating a passenger and freight service from its harbour to Waterford in Ireland, using the SS *Magnetic,* owned by the Waterford & Aberdovey Shipping Co – even cattle was allocated space on this ship.

As the line from Aberdovey to Llwyngwril was complete and opened on 24 October 1863, the station at Aberdovey was a terminus for trains from the north for almost four years, and the first train was ferried across the River Dovey from Ynys Las to Aberdovey.

From here, the 1863-built line to Towyn and Llwyngwril passed behind the sandy beach, overlooked by the hills above Aberdovey known as Cefn Rhos. On approaching Towyn (now named Tywyn), the Talyllyn Railway butts up to the main line on the right as coastal route standard-gauge trains enter Towyn station. This narrow-gauge (2ft 3in) railway reached Towyn in late 1865.

Built for the transportation of slate from quarries along its route, the Talyllyn Railway, together with the Corris and Mawddwy railways, used Aberdovey harbour to ship its valuable commodity. The Talyllyn Railway opened to both passengers and freight in 1866.

At Towyn's Cambrian station the main station buildings were on the up side, and a goods shed was situated beyond the up platform on the Aberdovey side of the station. Opposite the goods shed was a fine timber-built signalbox, with a separate small platform for the handing over of the token to down trains.

As the line singled on its northbound journey, Tonfanau and Llangelynin Halt were reached. Tonfanau's station buildings were very basic, and at this location was a level crossing, and quarry sidings, and military sidings were nearby. Llangelynin Halt's platform was very short (75ft) with a small corrugated-iron hut, two lamps and a comparatively large station nameboard.

On reaching Llwyngwril, a camping coach could be found after World War 2, on a disused siding on the up side, one of many strategically placed at small seaside resorts throughout Britain at this time. Llwyngwril possessed a huge water tower and water column, situated at each end of the down platform.

Between here and Fairbourne was the Friog avalanche shelter, situated halfway down a steep cliff below the coast road. Here, owing to the severe curves, trains screeched and groaned as they crept around this precipitous terrain. This was followed by a down gradient into Fairbourne of 1 in 55. The station here comprised just one platform, with brick and stone-built buildings. From here, the 15in-gauge Fairbourne Railway ran to the Mawddach estuary, where a small ferryboat conveyed passengers to Barmouth. Another camping coach was sited at Fairbourne behind the platform.

Barmouth Junction, graced with an impressive station nameboard stating 'Barmouth Junction – Change for Dolgelley, Bala, Corwen, Llangollen, Ruabon and Chester', was the next station on the line – its name changed to Morfa Mawddach in 1960. The triangular station layout, comprising four platforms, also enclosed a small goods yard. Tokens for the Dolgelley branch were regularly handed over on the station platform, just beyond the large signalbox. Even as early as the 1930s, a camping coach was situated here – a six-wheeled clerestory. One of the provisos of spending a holiday in a camping coach after the war was that you were compelled to travel by rail to the location.

From Barmouth Junction/Morfa Mawddach, the line crossed the River Mawddach over the famous Barmouth Bridge and over the Old Chapel Viaduct into Barmouth.

The line from Llwyngwril to Barmouth was opened to traffic on 3 July 1865, but because of bridge construction over the estuary, it did not open to Barmouth Junction for passenger traffic until 3 June 1867 and on 10 October 1867 for freight, at the same time opening from Llwyngwril to Pwllheli. Barmouth station was the largest on this coastal stretch of line, housing double track and a bay at each end. There were excellent freight loading facilities and the station was centrally-placed in the town – a resort popular with holidaymakers since the late-1880s. A good variety of motive power could be enjoyed here, with trains joining the Cambrian system from Dolgelley and Ruabon.

Beyond Barmouth, huge boulders protected the line from the sea as far as Llanaber Halt, a distance of about a mile. As the line headed towards the next major station, Harlech, it passed through Talybont Halt, Dyffryn Ardudwy, Talwrn Bach Halt and Llanbedr & Pensarn, where a short wooden viaduct crossed a tidal inlet. The final stopping place before Harlech was Llandanwg Halt, situated at the top of an incline where a 15mph speed restriction existed owing to unstable clay cliffs.

Harlech, majestically overlooked by its castle (a favourite point from where to photograph trains in the station below), was approached over two level crossings, the signalbox being situated between them on the down side beside the second of the two. Harlech's fine station buildings graced the down platform, and on crossing the line by the lattice-style footbridge, a pagoda hut was found at the foot of the steps, on the up platform. A loop line passed behind the up platform and could handle a train of at least 40 wagons. Although the station was situated way below the town, there was a large residential area in the vicinity.

From Harlech, the route was comparatively level as far as Penrhyndeudraeth, and it passed through Ty Gwyn Halt, Talsarnau and Llandecwyn Halt. The two halts comprised very small platforms with roughly square corrugated-iron shelters. Talsanau housed its signal cabin on its only platform, on the up side of the line and, after World War 2, was another location for camping coach – No W9922W – where steps provided access from track-level.

As the line approached Penrhyndeudraeth from Llandecwyn, it passed over the River Dwyryd by means of a toll bridge, where motor vehicles ran parallel with the railway. The river is noted for its trout, salmon and sewin. A quite extensive freight yard was once located at Penrhyndeudraeth and its well-situated station buildings on its only platform included an interesting signal cabin, brick-built, with covered steps to shelter staff from the elements.

From Penrhyndeudraeth, the route, on heading west, passed beneath the Ffestiniog Railway at Minffordd, where the line peaked after a climb of 1 in 65. From Minffordd it was downhill at 1 in 50 towards Portmadoc for ½ mile and then, more or less, over flat terrain into the station there.

Portmadoc station was approached over two level crossings, with cattle pen sidings between, on the north side of the line. A water tower was erected behind the station buildings on the up side. A sawmill and timber shed, with their own sidings, were situated on the down side, behind the station, and the goods shed was beside the down platform, and a siding passed through it, alongside the platform.

The engine shed was sited on the up side, facing Pwllheli, at the end of the station, and in December 1947 had an allocation of one ex-Cambrian 0-6-0, one Collett 0-6-0 (No 2204), three 'Dean Goods' engines, two 'Dukedogs', and two 4500 class 2-6-2Ts one large and one small. The later shed was a two-road building, constructed in brick, with a slate roof (appropriately – the local product). The original 1867 shed was moved to Pwllheli, the new one appearing circa 1907.

On the east side of Portmadoc station the Cambrian line crossed two narrow gauge tramways. The Gorseddau Tramway came into being in the 1850s, carrying horse-drawn traffic. Locomotive haulage appeared on the line in 1875 after takeover by the Gorseddau Junction & Portmadoc Railways Co. A little further east, the Croesor Tramway crossed the Cambrian to terminate alongside the Festiniog Railway station. The Croesor Tramway opened to traffic in 1864 and was used by the Welsh Highland Railway from 1923. After a few years, the Welsh Highland line terminated north of the main-line crossing and passenger traffic ceased in 1936, with freight continuing until 1937.

Trains ran to Portmadoc from Caernarvon and Afon Wen a little before the Cambrian opened its line to traffic from Barmouth to Portmadoc and Pwllheli, in October 1867.

From Portmadoc, the line continued through Wern, where a siding and goods depot were sited, mainly for stone trains, and on to Black Rock Halt, closed for passenger use in 1976. At Criccieth, a stone-built goods shed was situated at the east end of the up platform. On the down platform, also at the eastern end, was a 22-lever signalbox. The ruined castle overlooked the entire scene.

After about three miles, Afon Wen was reached, where the LNWR line joined from Caernarvon. In fact, the first 13 chains of the branch line were Cambrian-owned. The long up platform at Afon Wen measured 1,233ft in length and the main station buildings were on the island platform. A turntable was sited behind the down platform and, at this very exposed site, two water towers were placed within yards of each other to cope with the extensive traffic, particularly during summer weekends. The line from here to Penychain, the next station on the line, was doubled in 1947.

Penychain was opened by the GWR in July 1933 for holiday traffic but the holiday camp nearby was not opened to the public until 1947, its incomplete site being taken over by the Navy during World War 2 – it had both up and down platforms. At Abererch the one-platform station housed a timber waiting shelter and, beside that, at the end of the platform, a slate-built crossing-keeper's house adjacent to the level crossing.

At Pwllheli (pronounced 'Pullth-helly' and meaning salt-water pool) the engine shed was eventually sited south of the line, roughly 30 chains from the station. The original station was situated about ½ mile from the town – approximately where the later engine shed was sited. When the new inner harbour opened, the railway was extended about ½ mile along an embankment built for the harbour works. The new station opened in 1909 and the old buildings were re-erected at Aberdovey.

Evacuation specials operated to Portmadoc and Pwllheli during the war, from Liverpool, and over the seven-day period of the evacuation programme, many scheduled services were cancelled. The specials to Portmadoc and Pwllheli were operated on the 5th, 6th and 7th day of the programme and ran from Liverpool Lime Street and Wavertree. On the first and second day of the programme, for Birkenhead evacuees, trains ran to Penrhyndeudraeth, Talsarnau, Barmouth and Towyn from Birkenhead, and from Rock Ferry to Criccieth.

From 11 March 1942, permission was granted to operate certain LMS types between Afon Wen and Pwllheli, subject to specified speed restrictions. These included 'uncoloured' Class 2 and 3 0-6-0 freight engines, Class 1 'yellow' 2-4-2 passenger tanks, certain Class 3 'yellow' and 'blue' 0-6-0 freight engines, and 'red' Class 3 0-6-0 freight tanks. From Afon Wen to Portmadoc, 'uncoloured' Class 2 0-6-2 and 0-6-0 freight tanks were permitted, together with 'yellow' Class 1 2-4-2 passenger tanks, and Class 3 2-6-2 passenger tanks and 0-6-0 freight locomotives.

7827
75027

82020

GOGARTH HALT

Above left:
Dovey Junction, on 29 August 1963, showing an Aberystwyth train, hauled by GWR 'Manor' No 7827 *Lydham Manor*, on the left, and a Pwllheli-bound train, with BR Standard 4-6-0 No 75027, on the right. Passengers are seen rushing to change trains. *Andrew Muckley*

Left:
Standard 2-6-2T No 82020 is pictured at Dovey Junction with a train for Pwllheli. These Swindon-designed locomotives, built in 1952, did not reach the Cambrian until 1958/59. *E. N. Kneale*

Below left:
Gogarth Halt, one of the halts constructed by the GWR after the Grouping, sees Standard 2-6-4T No 80097 with a Machynlleth to Barmouth train in August 1964. Note the basic passenger shelter. *Bernard Robinson*

Above:
Single-chimney Class 4 BR Standard 4-6-0 No 75014 emerges from one of the tunnels on the line beside the Mawddach estuary, between Machynlleth and Aberdovey with a summer Saturdays Only working from London to Pwllheli.
N. B. Rutherford

Below:
The 10.55am Dovey Junction to Barmouth train enters Aberdovey station, on 27 August 1937, hauled by 'Earl' No 3204. The ex-Cambrian home signal adds interest to the scene. *LCGB/Ken Nunn Collection*

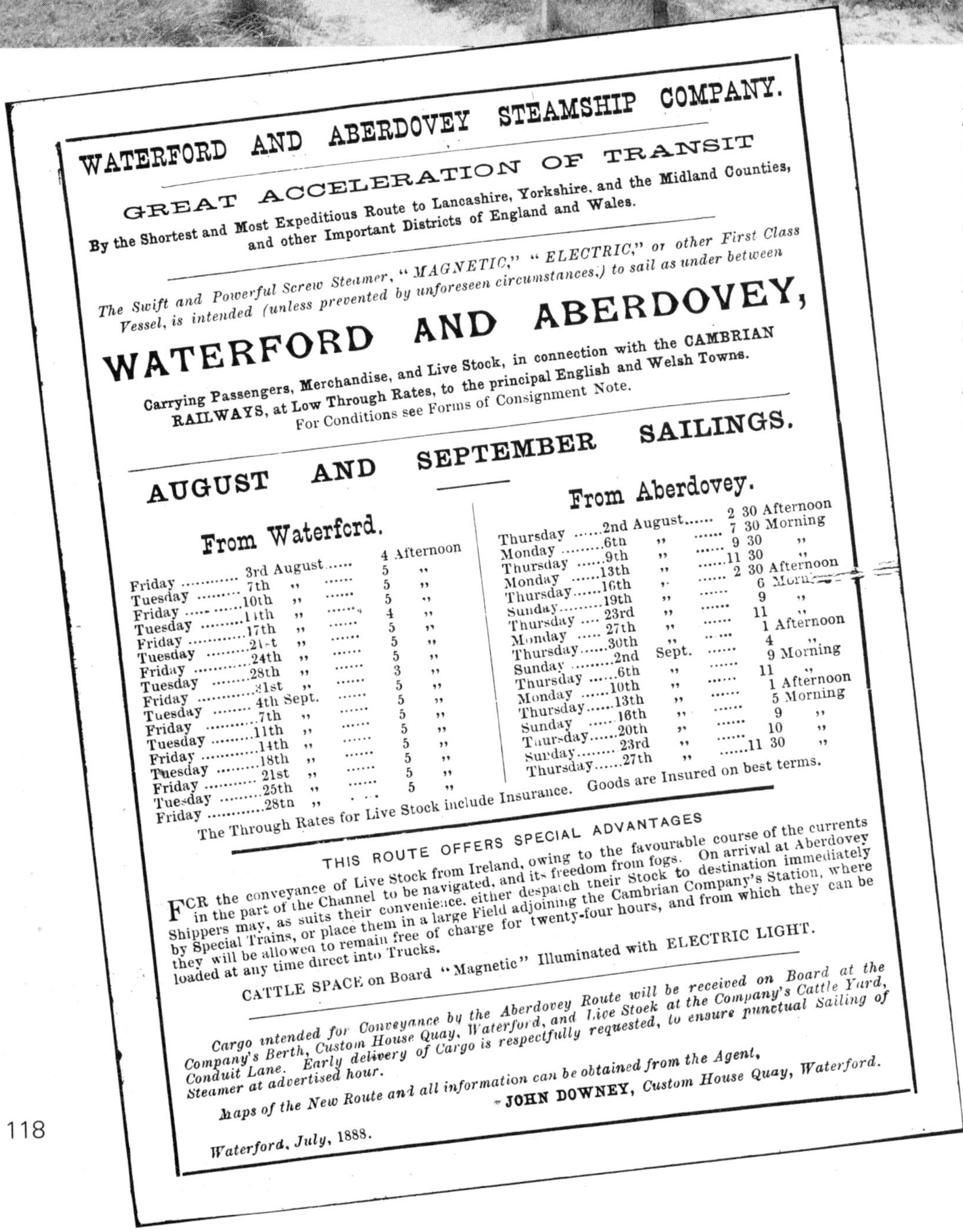

WATERFORD AND ABERDOVEY STEAMSHIP COMPANY.

GREAT ACCELERATION OF TRANSIT

By the Shortest and Most Expeditious Route to Lancashire, Yorkshire, and the Midland Counties, and other Important Districts of England and Wales.

The Swift and Powerful Screw Steamer, "MAGNETIC," "ELECTRIC," or other First Class Vessel, is intended (unless prevented by unforeseen circumstances,) to sail as under between

WATERFORD AND ABERDOVEY,

Carrying Passengers, Merchandise, and Live Stock, in connection with the CAMBRIAN RAILWAYS, at Low Through Rates, to the principal English and Welsh Towns.
For Conditions see Forms of Consignment Note.

AUGUST AND SEPTEMBER SAILINGS.

From Waterford.

Friday	3rd August	4 Afternoon
Tuesday	7th "	5 "
Friday	10th "	5 "
Tuesday	14th "	5 "
Friday	17th "	4 "
Tuesday	21st "	5 "
Friday	24th "	5 "
Tuesday	28th "	5 "
Friday	31st "	3 "
Tuesday	4th Sept.	5 "
Friday	7th "	5 "
Tuesday	11th "	5 "
Friday	14th "	5 "
Tuesday	18th "	5 "
Friday	21st "	5 "
Tuesday	25th "	5 "
Friday	28th "	5 "

From Aberdovey.

Thursday	2nd August	2 30 Afternoon
Monday	6th "	7 30 Morning
Thursday	9th "	9 30 "
Monday	13th "	11 30 "
Thursday	16th "	2 30 Afternoon
Sunday	19th "	6 Morning
Thursday	23rd "	9 "
Monday	27th "	11 "
Thursday	30th "	1 Afternoon
Sunday	2nd Sept.	4 "
Thursday	6th "	9 Morning
Monday	10th "	11 "
Thursday	13th "	1 Afternoon
Sunday	16th "	5 Morning
Thursday	20th "	9 "
Sunday	23rd "	10 "
Thursday	27th "	11 30 "

The Through Rates for Live Stock include Insurance. Goods are Insured on best terms.

THIS ROUTE OFFERS SPECIAL ADVANTAGES

FOR the conveyance of Live Stock from Ireland, owing to the favourable course of the currents in the part of the Channel to be navigated, and its freedom from fogs. On arrival at Aberdovey Shippers may, as suits their convenience, either despatch their Stock to destination immediately by Special Trains, or place them in a large Field adjoining the Cambrian Company's Station, where they will be allowed to remain free of charge for twenty-four hours, and from which they can be loaded at any time direct into Trucks.

CATTLE SPACE on Board "Magnetic" Illuminated with ELECTRIC LIGHT.

Cargo intended for Conveyance by the Aberdovey Route will be received on Board at the Company's Berth, Custom House Quay, Waterford, and Live Stock at the Company's Cattle Yard, Conduit Lane. Early delivery of Cargo is respectfully requested, to ensure punctual Sailing of Steamer at advertised hour.

Maps of the New Route and all information can be obtained from the Agent,
JOHN DOWNEY, *Custom House Quay, Waterford.*

Waterford, July, 1888.

Above:
Aberdovey, showing the trackbed of the old Harbour branch in the foreground. BR Standard 4-6-0 No 75063 heads a southbound freight on 16 July 1965. *C. L. Caddy*

Left:
The August/September 1888 sailings timetable of the Waterford & Aberdovey Steamship Company — using the SS *Magnetic*.
Author's Collection

Towyn

Above:
One of the unusual double-heading combinations found on the Cambrian. A 'Dukedog' heads a 4500 class tank on a passenger working, seen entering Towyn from the Aberdovey direction. Near this point, on the opposite side of the track, the narrow gauge Talyllyn Railway terminates. *N. E. Stead*

Below:
In July 1964, the daily up coast line pick-up freight, hauled by 'Manor' No 7823 *Hook Norton Manor*, carries out shunting duties at Towyn, before continuing on its journey south. *J. I. C. Boyd*

Top:
BR Standard 4-6-0 No 75002 pulls out of Towyn with a Barmouth to Machynlleth stopping train on 1 September 1964. The station area with its goods shed and signalbox can be seen in the background. *W. L. Underhay*

Above:
In April 1966, a special working arrives at Towyn, in preparation for Sunday work in strengthening the coastal defences of the town. In charge is BR Standard 2-6-0 No 78062. There was a remarkable likeness between these Derby-built Moguls and the Ivatt 2-6-0s in the 46400 series. *J. I. C. Boyd*

Above right:
With the Friog avalanche shelter in the background, and the sea below, Machynlleth-allocated Mogul No 6335 prepares to drift down-gradient from Friog into Fairbourne with a Pwllheli-bound passenger working in August 1958.
J. I. C. Boyd

Right:
This August 1957 view of Barmouth Junction shows 'Dukedog' No 9017 coupled to a 4500 class 2-6-2T having turned on the triangle in preparation to return to Barmouth to take over the down 'Cambrian Coast Express'. *G. F. Bannister*

BARMOUTH JUNCTION SIGNAL BOX
9017

Right:
After the change of station name from Barmouth Junction to Morfa Mawddach, in 1960, GWR 'Manor' No 7802 *Bradley Manor* pulls away with a summer Saturday special from Birmingham to Barmouth in August 1963.
Bernard Robinson

Centre right:
BR Standard 4-6-0 No 75051 is pictured at Barmouth Junction on 30 July 1958 with a 'Cambrian Radio Cruise' train. The train has paused to change tablets before crossing the Barmouth Bridge. The signalman, having completed the tablet exchange, can be seen returning to his box, across the Dolgelley line. *P. H. Wells*

Below:
The Royal Train (empty stock) heads south from Barmouth Bridge en route to Aberdovey in August 1963. 'Manors' Nos 7819 *Hinton Manor* and 7822 *Foxcote Manor* (both now preserved) have been delegated the duty.
Bernard Robinson

Left:
With sand blowing furiously, caused by the draught from the passing train, 'Dukedog' No 9014 crosses the embankment south of Barmouth Bridge with a Pwllheli to Machynlleth stopping train, on 24 July 1956. No 9014, in 1949, was fully lined-out, together with No 9009, but other members of the class carried a plain black livery. *R. E. Vincent*

Below:
With Cader Idris in the background, an Ivatt 2-6-0, having just left Morfa Mawddach, makes its way across the Mawddach estuary in January 1964 with a northbound train.
J. I. C. Boyd

·75009·

Left:
Barmouth Bridge over the Mawddach estuary, with Standard 4-6-0 No 75013 heading the up 'Cambrian Coast Express' on 19 July 1966. Barmouth is seen in the background as the train approaches the two-way distant signal. *M. Dunnett*

Below left:
Crossing Barmouth Bridge, in August 1963, is Standard 4-6-0 No 75009 with a train for Machynlleth. Apart from the 'Manors', these were the only 4-6-0s used on the Cambrian Coast line. *Bernard Robinson*

Above:
This fine view shows the entire length of the bridge over the Mawddach estuary, with Cader Idris standing majestically in the background. BR Class 4 Standard 2-6-0 No 76040 approaches Barmouth with a passenger train on 9 July 1966. Very few locomotives of this class were seen on the Cambrian. *C. L. Caddy*

Below:
Hauling a mixed freight, including some wagons of ballast, 4500 class 2-6-2T No 5541 enters Barmouth with a down train on 28 August 1955. From No 4575 onwards, locomotives in this class had larger side tanks with sloping tops with a greater water-carrying capacity. *P. H. Wells*

Barmouth

Above:
On 13 June 1957, GWR Mogul No 5399, with the Pwllheli to Ruabon afternoon local, passes 0-4-2T No 5809 at Barmouth station. The 'Coffeepot' is the Dolgelley auto-train engine. No 5399 was a Chester-allocated engine, all Cambrian Moguls being in the 6300 series. *G. F. Bannister*

Below:
Complete with polished copper chimney cap, 'Dukedog' No 9021 prepares to leave Barmouth with a down express for Pwllheli on 5 June 1955. The footbridge in the background served both passengers changing platforms and also pedestrians wishing to cross the line when the level crossing gates were closed. *P. H. Wells*

The 1930s

Above:
A good variety of motive power could always be found at Barmouth over the years and, hauling interesting coaching stock, 2600 class 'Aberdare' No 2679 prepares to leave the station with a Ruabon local working (c1932). These Dean-built engines were predominantly used on coal trains, and some were fitted with the larger ROD tenders. No 2679 was withdrawn in February 1946. *R. S. Carpenter Collection*

Below:
An early view of a 2251 class Collett locomotive, No 2270, with the 2.35pm train for Birkenhead, seen leaving Barmouth on 30 August 1937. During this time, coach roof boards were commonly used and, of course, smokebox numberplates did not come in until 1948. It was 1938 before members of this class were allocated to the Cambrian depots, to replace the 'Dean Goods' engines. *LCGB/Ken Nunn Collection*

The 1950s

Above:
Barmouth station in the 1950s, and Beyer Peacock 0-6-0 No 864 (ex-Cambrian No 38) awaits departure. A Belpaire 15 class goods engine, No 864 was not withdrawn until November 1952, and it is believed never to have carried smokebox numberplates. *H. N. James*

Right:
On 28 August 1952, 'Dean Goods' engine No 2323, allocated to Machynlleth, is seen with a local passenger working at Barmouth.
H. N. James

Below right:
Arriving at Barmouth from Dovey Junction, on 23 May 1955, is 2-6-2T No 5517, piloted by GWR 'Dukedog' No 9014, after a spot of bother. *P. H. Wells*

'Coffeepots' on Dolgelley Services

Right:
The Dolgelley auto-train at Barmouth, c1936 with 0-4-2T No 4867 in charge. In 1946, the entire 4800 class of 0-4-2Ts were renumbered in the 1400 series, leaving the 4800 series of numbers vacant for oil-burning 2-8-0s of the 2800 class — a short-lived conversion. No 4867 displays its GWR roundel on the tankside.
R. S. Carpenter Collection

Below right:
Over 20 years later, on 13 July 1957, the Dolgelley auto-train is hauled by 0-4-2T No 5801, still in GWR livery after nine years of nationalisation.
G. F. Bannister

MONA CAFE
CAFE GRILL ROOMS
RMO
GWR

Above:
Ivatt 2-6-0 No 46523 prepares to leave Barmouth with a two-coach Dolgelley train in May 1963. The water column, regularly used, is well illustrated. *C. L. Caddy*

Right:
Leaving Barmouth for Pwllheli, GWR 'Duke' No 3254 (9054) *Cornubia* heads north in August 1946 (during the period when the class was being renumbered). In the background are situated cattle pens and the goods shed. *John How*

Above:
BR Standard 2-6-2T No 82006 leaves Barmouth with a northbound train in August 1964. The large boulders on the right protect the railway (hopefully) from high seas. This 2-6-2T was soon to be transferred to Nine Elms on the Southern Region. *Bernard Robinson*

Below:
With a 'Cambrian Radio Cruise', 4-6-0 Standard No 75053 passes Talybont Halt, north of Barmouth. The 'Cambrian Radio Cruises' commenced in the summer of 1956 and were operated daily by the London Midland Region from Llandudno. Their route took them through North Wales and along the Cambrian Coast line and incorporated a 2hr stop at Aberdovey. The coaching stock used had individual armchair seating and a cafeteria was on board. A radio commentary, describing the scenery en route, was included in the tour. *P. Poulter*

Above:
The headboard used on the 'Cambrian Radio Cruise' *Author's Collection*

Below:
BR Standard 2-6-4T No 80098 prepares to leave Dyffryn Ardudwy with a northbound stopping train in August 1964. The brick-built station buildings can be seen on the up platform. *Bernard Robinson*

Above right:
Collett 0-6-0 No 2202, carrying the official crested 'Cambrian Coast Express' headboard, awaits the right-away from Llanbedr & Pensarn station on Bank Holiday Monday, 4 August 1958. This train is the up Criccieth to Paddington portion of the 'CCE'. *J. J. Holden*

Below right:
In July 1959, a local passenger working from Machynlleth to Pwllheli draws into Llandanwg Halt. This station was built by the GWR in 1929. No 82000, the BR Standard 2-6-2T in charge, was built at Swindon in the first of five batches and introduced in 1952. A further order, in 1954, increasing the total in the class from 45 to 63 was cancelled, owing to the advance of dieselisation. *G. F. Bannister*

CAMBRIAN COAST
EXPRESS
2202

82000

Harlech

Above:
Harlech, on 24 August 1961, and 2251 class Collett 0-6-0 No 2294 potters through the station with a mixed freight. The lattice footbridge and basic accommodation on the up platform are clearly visible. *A. Tyson*

Below:
This view of double-headed 4500 class 2-6-2Ts, with a passenger working at Harlech, on 13 September 1949, shows the differing styles of this class. The rectangular tank on the leading engine indicates that it is in the 4500-4574 series, whereas the sloping larger tank on the second engine shows that it is one of the later members of the class, capable of carrying 300gal more water, making the locomotive four tons heavier. *Oakwood Collection*

Above:
On 13 September 1949, 4-4-0 'Dukedog' No 9000 stands in the loop which passes behind the up platform at Harlech, and beside the temporary Lever's cattle food depot, a clerestory coach, eventually being replaced by a modern building. *Oakwood Collection*

Left:
The basic Ty Gwyn Halt, in the 1950s, as 2251 class 0-6-0 No 2202 passes with a mixed freight train. The duties for this class varied from the 'Cambrian Coast Express' to freight and engineer's trains.
G. F. Bannister

Right & Below right:
Crewe South-allocated Ivatt 2-6-0 No 43151 hauls the inspection saloon and passes Llandecwyn Halt on 11 July 1966. In the first view, the very basic platform at this location can be seen, and the other view shows the inspection saloon from the rear and the bridge over the River Dwyryd as the line approaches the toll bridge near Penrhyndeudraeth. These Class 4 Ivatt Moguls were 12 tons heavier than the Class 2 46400 variety.
C. L. Caddy

Below:
BR Standard 4-6-0 No 75047 crosses the Afon Glaslyn, between Portmadoc and Minffordd, with the 12.50pm Pwllheli to Machynlleth pick-up freight, on 29 July 1966.
J. R. P. Hunt

Afon Wen

Left:
Afon Wen station, junction with the L&NWR line from Caernarvon. In this view, 2251 class 0-6-0 No 2294 prepares to leave for Portmadoc with the 1.30pm train from Pwllheli on 28 August 1961. *A. Tyson*

Below left:
On 11 April 1939, Great Western and LMS motive power are seen together at Afon Wen. A 4500 class 2-6-2T passes an LMS push-pull working propelled from the centre by a Class 3MT Fowler 2-6-2T. *H. N. James*

Below:
This view of Afon Wen clearly shows the station layout, as 'Mickey Mouse' No 46428 takes on water before continuing its journey. A feature of this station was the two steel footbridges situated at each end. The main station buildings were sited on the island platform. *Oakwood Collection*

Pwllheli

Above:
In April 1963, a Standard 2-6-2T, No 82021, of a class extensively used in the area, prepares to leave Pwllheli with a two-coach train. Like many seaside termini, Pwllheli was quite an impressive station in its day. *Andrew Muckley*

Below:
The two-road Pwllheli engine shed which replaced the single-road depot in the mid-1950s. The two-road shed was, in fact, the third built at Pwllheli, the original single-road building being situated on the other side of the line, about 10 chains further east. BR Standard 2-6-2T No 82000 and Collett 0-6-0 No 3208 are pictured on shed. The Pwllheli-allocated Collett 0-6-0s, prior to 1957, were used on the Western Region's 'Cambrian Land Cruise' trains. The route of this train was Pwllheli-Barmouth-Corwen-Rhyl-Bangor, and back to Pwllheli. However, from the summer of 1957, this train was increased to seven coaches, which was above the permitted load for the Collett 0-6-0s. To solve the problem, a BR Standard 4-6-0 was borrowed from Oswestry for this working, which ran on Tuesdays and Thursdays (Wednesdays in August). *E. N. Kneale*

15:

Barmouth Junction to Dolgelley

The first section of this line to open terminated at Penmaenpool. This opened to the traffic on 3 July 1865 — it continued to Dolgelley in 1869, opening to passengers on 21 June and to freight on 1 October. However, the passenger route was not connected to the Great Western Railway's station at Dolgelley until 1 August 1869, but Cambrian territory ceased 30 chains west of the station.

The line's completion to Barmouth Junction provided an alternative route between Merseyside and the Welsh coast, without having to pass through Caernarvon to Afon Wen.

In March 1870 a single platform was constructed at Arthog and 1½ miles from there, on the Penmaenpool side, a narrow-gauge tramway once existed to Tyn-y-Coed quarry.

In 1888, trains took about 25min to cover the line between Dolgelley and Barmouth and the service consisted of seven passenger trains and two freight trains in each direction. By 1923 passenger workings in each direction had been increased to 11 — two being rail motor car services (one class only). The journey time did not alter.

Torrential rain closed the line on 12/13 January 1965, prior to its officially announced closure date of 16 January. The line was breached each side of Dolgelley, and half a mile of track was washed away. This followed eight days of continuous rainfall and, as an example, Blaenau Festiniog recorded 18½in over that period, 5.14in falling on the last day. The line was also breached two miles east of Bala, on the Ruabon line.

From 1863 to 1865, Sharp Stewart built 12 2-4-0 passenger engines (all originally named) and No 56 (originally *Whittington*) worked from Penmaenpool. This locomotive was converted to a 2-4-0T in May 1907 and it was allocated the number 1191 by the GWR — it was withdrawn in 1922. In fact, none of the six locomotives of this class which passed to the GWR ever carried their GWR numbers. The bogie side tanks built by Nasmyth, Wilson in 1895 were also allocated to the line's engine shed, Penmaenpool, but not at first. They worked through trains from the GWR line.

A clearance test was made from Ruabon to Barmouth by GWR 2-8-0 No 2853 in August 1929, which stopped short of the Mawddach estuary viaduct, owing to previous restrictions banning the crossing of heavy locomotives. However, permission was granted as far as Barmouth Junction, but then 2-8-0s had to be turned on the triangle there. In December 1935, GWR 2-6-2Ts (Classes 5100 and 6100) were passed to work between Ruabon to Barmouth Junction.

At Penmaenpool, the only shed on the Cambrian section of the route to Dolgelley, which opened in 1869, the allocation at December 1947 was just one engine — GWR 0-4-2T No 1465, but in latter years Standards were regular visitors to this location.

Its original depot building was reclad in corrugated-iron sheets in Great Western days. It was a two-road shed measuring 60ft × 30ft.

Apart from Arthog, Penmaenpool was the only other intermediate station on the line. At this point a toll bridge crossed the River Mawddach, shortening the road journey to Barmouth, and avoiding travelling into Dolgelley. The George III hotel has now taken over some of the original station buildings at Penmaenpool.

In the days when ocean-going ships set out from this location, ships were actually built nearby. The signalbox still remains as an information centre for the RSPB.

As the line approached Dolgelley, about two miles from Penmaenpool, the River Mawddach appeared on the south of the line, which had crossed the river a mile out of town. The station area at Dolgelley has now been swallowed up by a bypass.

In its final days, the line was worked by Standard 4-6-0 75000, Ivatt 2-6-0 Moguls, and the lighter 2-6-0 Standards (at 86 tons), the 78000 class, which resembled Ivatt's 2-6-0s.

Left:
2251 class 0-6-0, No 3205, exchanges tablets at Morfa Mawddach, on 29 August 1962, before completing the last stage of its journey with the 7.17am train from Ruabon to Barmouth. *Ian Allan Library*

Passing Morfa Mawddach signalbox, BR Class 4 No 75009 takes the Ruabon line with the 2.55pm Barmouth to Chester working, on 31 October 1964. *J. C. Haydon*

Ruabon Trains

Left:
On 12 August 1959, GWR Mogul No 6320 awaits the right-away at Barmouth Junction with an afternoon Barmouth to Ruabon local train. This was, in fact, a Westbury-allocated locomotive, and was possibly on a running-in turn after works overhaul. *G. F. Bannister*

Left:
Another GWR Mogul, No 6324, enters Morfa Mawddach station with a train from Ruabon on 7 August 1956. On this occasion, the Mogul was allocated to Wolverhampton (Oxley). *G. F. Bannister*

Below:
Arthog station, on 27 August 1963, and a local Barmouth to Dolgelley train, hauled by 'Mickey Mouse' No 46521, is seen arriving. This particular engine was a regular on this branch. *Andrew Muckley*

Above:
The two-road shed at Penmaenpool could house one engine on each line. The shed was situated on the south side of the branch line, seen going away towards Morfa Mawddach, and 2-6-0 No 46521 and 4-6-0 No 75029 were resident on 12 July 1964. *C. L. Caddy*

Below:
The Dolgelley branch ran beside the River Mawddach and beneath the mountainside which rose to Cader Idris, at Penmaenpool. The George Hotel sat beside the track, between the station and the engine shed. Ivatt 2-6-0 No 46446 is seen with its two-coach branch-line train, on 3 June 1964, forming a Barmouth to Dolgelley local. *Derek Cross*

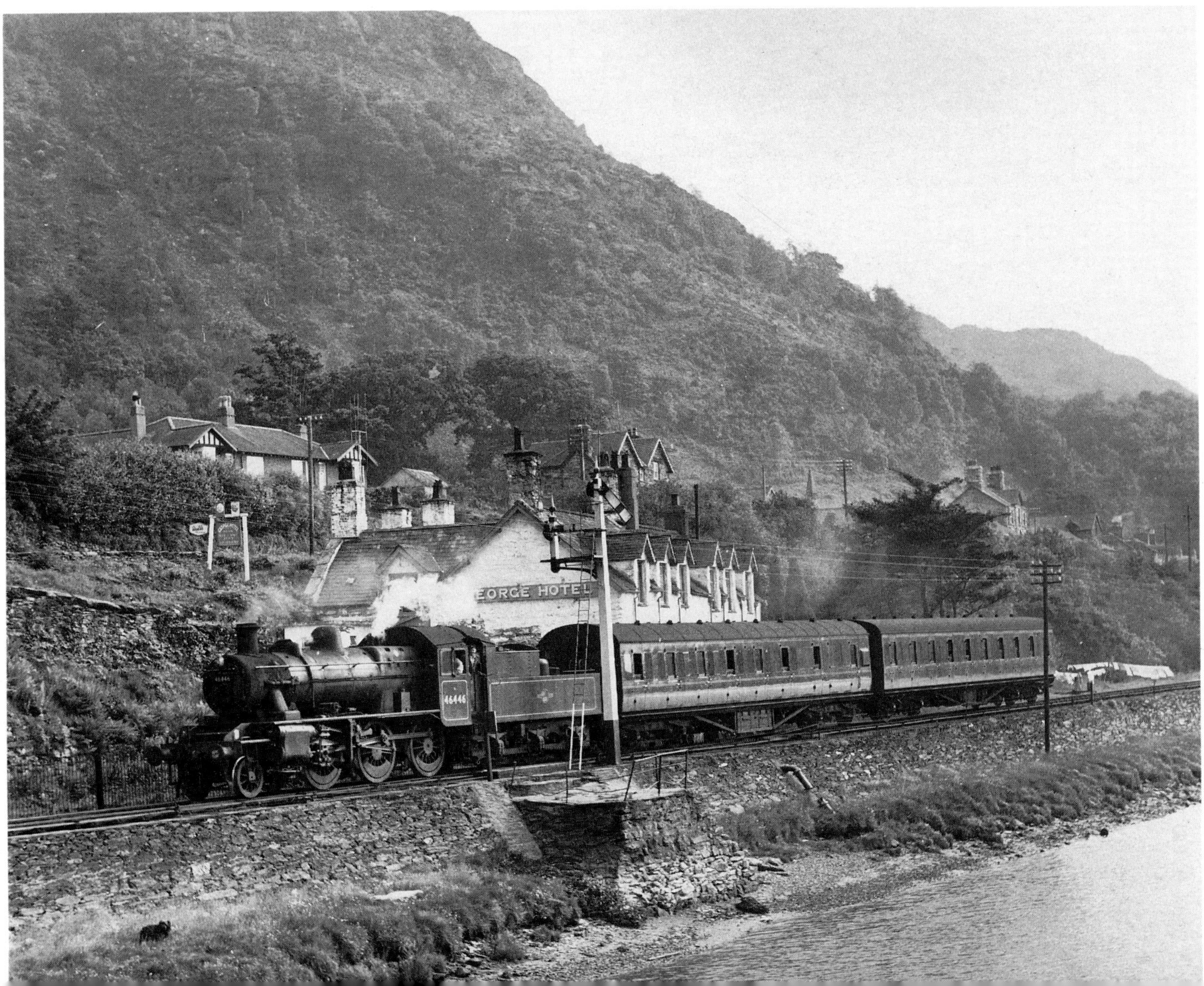

Above:
The boarded platforms at Penmaenpool provided very basic passenger shelter and the (now preserved) signalbox was situated at the end of the platform, near the toll bridge which took motor vehicles across the River Mawddach to the Dolgelley to Barmouth road. No 46521 enters the station on 27 August 1963 with a Barmouth to Dolgelley working.
Andrew Muckley

Below:
In this view, the small siding, situated on the roadside of the station at Penmaenpool, is seen as Ivatt 2-6-0 No 46521 stops with a Dolgelley train from Barmouth in August 1964.
Bernard Robinson

Above:
A BR Class 2 Standard Mogul, No 78002, is seen entering Dolgelley from Morfa Mawddach with a freight as it skirts the River Mawddach which it has just crossed. *C. L. Caddy*

Dolgelley

Below:
The station at Dolgelley itself was GWR-owned and, having just entered 'Cambrian' territory, Ivatt Mogul No 46509 prepares to continue its journey to Barmouth in August 1964.
Bernard Robinson